John Anderson

A Contribution to the Herpetology of Arabia

With a Preliminary List of the Reptiles and Batrachians of Egypt

John Anderson

A Contribution to the Herpetology of Arabia
With a Preliminary List of the Reptiles and Batrachians of Egypt

ISBN/EAN: 9783744788168

Printed in Europe, USA, Canada, Australia, Japan

Cover: Foto ©berggeist007 / pixelio.de

More available books at **www.hansebooks.com**

A CONTRIBUTION

TO THE

HERPETOLOGY

OF

ARABIA.

WITH A PRELIMINARY LIST OF THE

REPTILES AND BATRACHIANS

OF

EGYPT.

BY

JOHN ANDERSON, M.D., LL.D., F.R.S.

LONDON:
R. H. PORTER, 7 PRINCES STREET, CAVENDISH SQUARE.
1896.

ALERE FLAMMAM.

PRINTED BY TAYLOR AND FRANCIS,
RED LION COURT, FLEET STREET.

CONTENTS.

PART I.

A SKETCH OF THE PHYSICAL FEATURES OF THE COAST

OF

SOUTH-EAST ARABIA

AND OF THE

COUNTRY BETWEEN MAKALLAH AND THE HADRAMUT.

THE first part of a valuable memoir on the South-East Coast of
Arabia, from the entrance to the Red Sea eastwards to Misenát
in 50° 43′ 25″ E. longitude and 15° 3′ N. latitude, by Captain
Stafford Bettesworth Haines, of the Indian Navy, was read before
the Royal Geographical Society on the 11th May, 1839[1]. The
second part of the memoir did not appear until 1845[2]. It carries
the Survey as far east as Rás Jezirah, known as Cape Isolette,
but more properly Cape Island.

The Survey of this coast was completed by the Indian Navy
between 1844–46. It was under the direction of Captain
Saunders and Lieutenant Grieve, and was conducted from east
to west, beginning at Maskat. Only a short memoir of this
Survey was published by Commander Saunders[3]; but the late
Mr. H. J. Carter, F.R.S., the Surgeon of the surveying ship, the
'Palinurus,' gave an account[4] of his own observations, in order
to complete the geographical description of the coast. Separate
contributions to our knowledge of some parts of the coast-line
were made by other officers of both surveys.

[1] Journ. Roy. Geogr. Soc. vol. ix. 1839, pp. 125–156 and map.

[2] Ibid. vol. xv. 1845, pp. 104–160 and map.

[3] Ibid. vol. xvi. 1846, pp. 169–186.

[4] Journ. Bombay Branch Roy. As. Soc. iii. Part ii. 1851, pp. 224–317.
I am indebted to Mrs. Carter for the use of a reprint of this paper, illustrated
by Mr. Carter's sketches of the country and its people.

As Captain Haines and Mr. Carter did not confine their researches merely to the coast, but gave an insight into the character of the country lying beyond it, I have thought the subject of sufficient interest, in view of the zoological collections made on Mr. Bent's Expedition to the Hadramut being the first that have been obtained from South-Eastern Arabia, to justify my giving a brief summary of the leading features of the country between Aden and Rás el Had, and a sketch of Wrede's [1], Hirsch's [2], and Bent's [3] impressions of the Hadramut itself.

Round the headland of Jebel Shamshan, on which Aden is situated, lies the great bay of Ghubbet Seïlán, from which a plain extends into the interior. This plain was traversed by Captain S. B. Miles and M. Munzinger in 1870 [4]. They went to Bir Ali, 220 miles to the east of Aden, in a small *sambuk*, and thence penetrated into the interior as far as Habban and across the plain to Aden, through the country occupied by the Fudhlí tribe. The plain is about 200 square miles in extent, and is watered by two rivers, the Hassán and Banna; and when Captain Miles crossed the latter in the end of July, he says it was 400 [5] yards broad, and running over knee-deep. Along the shore the plain was bordered by a thick forest of acacia, and towards the hills broad fields of grass and corn stretched away to the Yaffai valley [6]. The uncultivated parts were either sandy patches, or were covered with brushwood and thick jungle. The jowárí grew to a great height, considerably overtopping the

---

[1] "An Excursion in Hadramaut by Adolph, Baron Wrede," Journ. Roy. Geogr. Soc. xiv. 1844, pp. 107-112; 'Reise in Hadhramaut,' edited by H. F. von Maltzan, 1870.

[2] Verh. Ges. für Erdk. Berlin, xxi. 1894, pp. 126-136 and map.

[3] Geogr. Journ. iv. 1894, p. 315 and map.

[4] Proc. Roy. Geogr. Soc. xv. 1871, pp. 319-328; Trans. Bombay Geogr. Soc. xix. 1874, pp. 166-186. Accompanying this summary of the Narrative issued by the Government of Bombay is a paper by M. Munzinger on the geographical features, geology, and hydrology of the triangle between Aïn Jowárí, Habban, and Haurá.

[5] This is probably a misprint for 40 yards.

[6] I cannot refrain from calling attention here to a statement by J. P. Malcolmson, in his account of Aden (Journ. Roy. As. Soc. viii. 1846, pp. 279-292), that a few hyænas of small size occur in the deep ravines inland from Aden. When at Suakin I was told that a small hyæna frequented the plain near that town. It proved to be not a hyæna, but *Proteles cristatus*, Is Geoffr. Is it possible that the small hyæna of the Aden ravines is the same animal?

head of a man on a camel. The hills are stated to abound
with myrrh trees.

About thirty miles inland rises the high mountain-range called
Jebel Yaffai, attaining to an elevation of over 6000 feet above
the sea. It has numerous fertile valleys that produce coffee,
dúra, and other crops. Rás Seïlán, the eastern extremity of the
bay, is a low sandy point on which a few date trees grow.
Ten miles to the east of this cape, and about two miles from the
coast, the country is well watered and cultivated; but beyond
the fact that partridges are found on it, nothing is known of
its fauna. At the village of Sughrá good water, bullocks, sheep,
poultry, onions, and pumpkins were easily procured. Sixteen
miles to the north-east of Sughrá, Jebel Kharaz towers to a
height of 5400 feet above the sea, and has the Wádí Bahreïn
winding through it, abundantly supplied with streams flowing
into an extensive lake which gives its name to the valley. Then
follows a tract of low, barren, sandy coast, succeeded by a range
of limestone mountains, twenty miles in length, and within five
miles of the sea, with its summits broken up into peaks and bluff
points. Further on, a number of black hills and rocky points
occur at intervals, close to the sea, and then follows a long stretch
of low sandy coast with more rocky points until the town of
Howaiyeh is reached, five miles inland, and situated on a wide
plain, the inhabitants of which were chiefly employed in agri-
culture. Here the surveying-ship the 'Palinurus' secured some
fine bullocks, good water, and excellent fish. Inland from Jebel
Makánátí, four miles north-east of Haurá, is the entrance to
the Wádí Meïfah, one of the great valleys of the coast, in a
prolongation of which lies the remarkable ruin, Nakab el
Hajar, visited by Lieut. Wellsted[1] and Mr. Cruttenden in
April 1835.

Landing from the ship, they crossed a belt of low barren sand-
hills and passed the two villages Aïn Abú' Ma'bad and Aïn Jowárí.
Continuing their way across a waste of low sandy billocks rising
in sharp ridges, followed by a sandy expanse covered with
stunted tamarisks which afforded a slight shade from the
scorching sun, they reached a tableland about 200 feet above the
surrounding plain, intersected by numerous ravines, the beds of

---

[1] Journ. Roy. Geogr. Soc. vii. 1837, pp. 20-34; 'Travels in Arabia,' 1838;
Haines, Journ. Roy. Geogr. Soc. ix. 1839, p. 143.

former torrents. The surface of the flat-topped hills was strewn with fragments of quartz and jasper. Leaving this barren plateau, they met with stunted acacias which increased in number as the travellers advanced; and then they came upon good water surrounded by trees, among which were numerous tamarisks and *Cissus arborea*, Forskål. The ground they next passed over afforded ample evidence of its having been quite recently the bed of a powerful stream. Numerous hamlets were seen among extensive groves of date-palms and verdant fields of dúra, and there were many herds of sleek cattle. On this, the second day of their journey, they travelled on till midnight, and, on the following morning, were astonished to find themselves surrounded by luxuriant fields of dúra and tobacco, extending as far as the eye could reach, mingled with the foliage of the acacia and the stately date-palm. The creaking of numerous wheels for the irrigation of the fields, several rude ploughs drawn by oxen, the ruddy countenances and lively appearance of the people, and the delightful refreshing coolness of the morning air, combined to form a scene which, Wellsted says, could never have been anticipated from the barren aspect of the coast where they had landed.

It was at this part of the coast (Bir Ali [1]) that Captain Miles and M. Munzinger entered the country in 1870. The town of Habban which they visited lies a considerable distance inland, situated in a gorge girt round, on every side, with high, almost inaccessible cliffs. It presents a striking appearance, as the houses are lofty, detached, castle-like structures. Around the town, wheat, jowárí, barley, and other crops are cultivated, and four crops are raised annually, viz., one rain-crop, and three by irrigation.

Near to Bir Ali is Hisn Ghoráb, a dreary-looking, brown hill, 464 feet high, in the neighbourhood of which the first Himyaritic inscription was found by Lieut. Wellsted and Dr. Hulton on 6th May, 1834. Close to this spot, a remarkable, flat-topped sandstone hill, called Sha'rán, rises from the plain to a height of 800 feet. Its summit is a crater-shaped cavity, 2500 yards in diameter, filled with salt water, and presenting the remarkable feature that the edge of the water is fringed by an overhanging

[1] Anthracite coal exists at Bir Ali, and specimens used to be taken to Aden as coal. Bitumen is found in abundance, and there are signs of copper.

bank of mangrove trees. This elevated crater-lake is called Kharif Sha'rán, and the view from it is described as both romantic and beautiful. Below the spectator are the dark waters of the crater with its fringe of trees, while, on one side, are rocky heights frowning over fertile valleys, and, on the other, the blue sea, with an island or two in the distance [1]. At the town of Kharïjáh, still further to the east, the country in places is again fertile, abounding in grass and date-palms, with excellent pasture-lands affording food to numerous herds of cattle; but, with the exception of these occasional oases, the coast-line is essentially barren.

Beyond Rás Rehmát the land is bold, with a succession of rocky points; but, a little to the east, the town of Al Ghaïdhar is embosomed in luxuriant groves of date-palms. Further on is the headland of Rás Barúm, with its valley of the same name, with palm trees, whilst the inland valleys here produce large quantities of dúra. The mountains that define them rise to an altitude varying from 5000 to 6000 feet; and their summits are said to be occasionally covered with snow in the cold season. Capt. Haines has stated, from personal observation, that heavy and continuous rains fall in November and December, July and August, and even in April and May; and he records that he has seen rain for three consecutive days.

From Barúm to Makallah, the coast is low, barren and sandy, forming a slight bay with great mountains in the background, chiefly composed of limestone, but with interbedded sandstones and masses of granite and basalt.

At Rá} Makallah the hills come down to the sea, and above the town they rise to about 300 feet as a reddish cliff, while above this towers the flat-topped summit of Jebel Gara. A few miles further on is the village of Bú Heïsh, surrounded by date-trees, in a well-watered valley about $1\frac{1}{2}$ mile from the shore. Another fertile district lies around the town Súku-l-Basír (the Ghaïl ba Wazír of Hirsch), a few miles north-west of Shehr on the coast. Sixty years ago, Capt. Haines found at the latter place much tobacco, plenty of vegetables, good dates, and pure water. Although other small oases are present, the coast-line from Makallah to the cliffs of Hámí, thirteen miles beyond Shehr, is

---

[1] I have consulted Capt. Haines's original MS. preserved in the India Office. It is illustrated by some sketches; and among them there is a pen-and-ink drawing of this lake taken from the margin of the crater.

an almost unbroken line of low barren sand, but the village of Hámí itself is situated in a picturesque ravine, with a grove of date-palms, and cultivated land near the beach. Capt. Haines and Lieut. Wellsted were the first to describe the hot springs in this part of the Arabian coast, to the presence of which the oases are largely attributable, combined with the drainage from the mountains that finds its way down the ravines on to the Sahil, or maritime plain. Capt. Haines ascertained that some of the springs had a temperature of 140° Fahr. Mr. Carter says they occur in such profusion between Makallah and Sihút, at the entrance of the Wádí Masílah, as to constitute one of the striking features of this part of the coast-line. The same traveller was also the first to call attention to another remarkable appearance presented by this plain, namely, the presence of extensive out-flows of basaltic rock, associated with volcanic cones rising to about 100 feet above the level of the ground. The basalt, from its blackness, is in strong contrast to the rest of the sandy Sahil, as a whole; and is so unmistakably volcanic, that but for its being unattended by any active signs of eruption, it might be mistaken for a recent lava outflow. These two features of the Makallah-Sihút Sahil could not fail to attract the attention of every traveller. They have recently been redescribed by Mr. Bent in his account of his visit to the Hadramut.

At Misenát, opposite to the opening of the Wádí Sheikháwí, the land is swampy and mangrove trees are numerous. The officers of the 'Palinurus' found, a little to the east of this valley, a number of Himyaritic characters in red paint, similar to those discovered at Hisn Ghoráb.

Immediately to the east of Sihút is the great opening of the Wádí Masílah, leading to the Hadramut, and the grandest of all the valleys that run inland and seem to divide the mountains of South Arabia into separate tracts. A few miles to the east of this valley rises the remarkable headland of Rás Sharwón, capped by two natural pillars seen at a distance of 60 or 70 miles; and further on lies the village of Hishn [1], described by Capt. Haines, and recently by Mr. Bent. Fifty miles further to the east rises the headland Rás Farták, and, next to Rás Seger, the boldest cape on this coast, and marking the boundary between the Mahrah and

---

[1] Niebuhr ('Descr. de l'Arabie,' 1774, p. 248, tab. xvii.) has given a plan of this port which he received from an Englishman he met in Bombay.

Gara tribes, which were described by Carter about fifty years ago [1]. This great promontory sweeps round to the east in one of the grandest escarpments on the coast. It is six miles in length, and, although quite perpendicular, is deeply worn into shelves under the shelter of which the people live; and as night comes on, the lights of these rock-dwellings are seen flickering on the face of the precipice. Mr. Carter observed the people moving about in the most perilous positions, and adds that in all probability the great size of the cliff rendered it difficult to form a just estimate of the width of the shelves; but the Mahrah pilot of the 'Palinurus' assured him that it was no uncommon occurrence for them to fall over and be drowned. This headland defines the western limit of the bay of El Kamar, inland from which runs another great level expanse, wholly barren with the exception of a few desert herbs. It is the beginning of another enormous valley, along which trade is said to be carried on with the Hadramut. The eastern side of the expanse terminates at Rás Tharbat Ali, 200 feet high, the seaward end of the Fatták ridge of mountains, immediately to the east of which lies another valley with the village of Damkot at its entrance, on a narrow sandy shore where a few miserable date-palms struggle for existence. This village is closed in, except towards the sea, by inaccessible mountains 3000 feet in height, perfectly barren, save on their summits which are more or less covered with grass and dotted over with small trees. The coast preserves this character as far as Rás Seger, a distance of about forty miles; but here and there a few narrow ravines lead down from the mountains. Carter visited one of these gorges, and found its sides wooded with acacias, balsams, and euphorbias.

Rás Seger is a gigantic headland, 3380 feet high, the sides of which, where not perpendicular, are covered with trees, and the plateau above with long grass, while numerous caves occur in the precipices. Beyond this headland is Rás el Ahmar, or the Red Cape, defining the western limit of the fertile maritime plain of Dhofár which is shut in behind by lofty mountains. It is the most favoured spot on the coast of South-East Arabia, and is the land of the famous frankincense tree.

[1] "Notes on the Mahrah tribe of Southern Arabia, with a vocabulary of their Language, to which are appended additional observations on the Gara tribe.' Journ. Bombay Branch Roy. As. Soc. ii. 1848, pp. 339-364. "Notes on the Gara tribe," *id. op. cit.* pp. 195-201, and plate.

This plain of Dhofár was explored by Mr. C. J. Cruttenden in 1834. It has been described by him[1], also by Capt. Haines[2], and by Mr. Carter. Mr. Cruttenden travelled over it on foot accompanied by two men of the Gara tribe. He describes its rich vegetation and that of the hills, and mentions the lime, tamarind, henna, nebbuck, tamarisk, *dom* (*Zizyphus spina-christi*), the *subhan* or frankincense tree, the abundance of aloes, and the figs and grapes of the higher region. The running streams, the large sheets of water on the plain, the flocks of sheep and goats, the ruins of El Balad, and the remarkable ravine of Darbat behind Takáb, are all enumerated; but unfortunately, like all the travellers that have followed him, he gives no information about tho wild animals of the country beyond stating that the only beast of prey on the plain of Dhofár is the hyæna, and that antelopes are numerous. Haines describes the plain as covered with large tracts of maize and millet, and the trees so abundant as to afford ample shade from the scorching rays of the sun; the whole being richly watered by streams from the mountains. The plain is 50 miles in length and 6 to 12 miles in breadth. The mountains approach it in sudden descents; and some of their ravines open on to it in abrupt precipices over which streams fall into the gorges below. One of the most striking of these ravines is that of Darbat, described fully by Carter[3] many years ago. He followed the Khor Reri, and, entering the bottom of that ravine, found it suddenly closed by a precipice 250 feet in height, and, scaling it, arrived at a grassy plateau shut in on every side by the mountains, except towards the sea, where it terminated in the precipice just mentioned. This sequestered hollow was occupied by a small lake and stream, which were diverted for the irrigation of crops of indigo, corn, and onions. The lake, on which water-fowl floated, was fringed in many places with tall bulrushes and spreading trees; and among them and on the slopes were pomegranate bushes and fig trees. The precipitous sur-

---

[1] Proc. Bombay Geogr. Soc. 1837-38, pp. 70-74; Trans. Bombay Geogr. Soc. i. 1844, pp. 184-188.

[2] Journ. Roy. Geogr. Soc. xv. 1845, pp. 116-122.

[3] Journ. Roy. Geogr. Soc. xvi. 1846, pp. 169-186; Journ. Bombay Branch Roy. As. Soc. iii. 1849-51, pp. 252-264. For a geological account of Dhofár, see a memoir on the Geology of the South-East Coast of Arabia, Journ. Bombay Branch Roy. As. Soc. iv. Jan. 1852, pp. 32-44.

roundings of this plateau were here and there perforated by deep caverns inhabited by Guras; and in one of them Carter spent the greater part of a day with the Sheikh who lived in it. Like those at Rás Fartak, these caverns were visible at night by their lights, to those on board the 'Palinurus.' Mr. Carter has not only given a full description of the physical characters of the plain and its mountains, but he has also described the inhabitants, the frankincense tree, and the ruins of El Balad [1].

The mountains lying behind the plain of Dhofár were all designated by Capt. Haines as the Subhán range, and in 1834 or 1835 Mr. J. Smith, purser of the 'Palinurus,' traversed these mountains in perfect safety, and, under the name of 'Ahmed,' became a great favourite with the inhabitants. He was everywhere hospitably received, and they would not allow him to drink water of the clear mountain-streams that were meandering in every direction: "No," they said, "do not return, Ahmed, and say that we gave you water, while our children drank nothing but milk." In every instance they gave him the warmest place by the fire, invariably appointed some one to attend to his wants, and even extended their hospitality so far as to offer him a wife and some sheep, if he would only stay and reside among them. On Mr. Smith expressing a wish to see some of the numerous wild animals, the footprints of which were everywhere visible, on what he describes as the park-like mountains, they immediately despatched a party, who returned with a splendid specimen of an ibex, a civet cat, and a fine ounce [2]. Mr. Smith himself saw plenty smaller game, such as antelopes, hares, foxes, guinea-fowl, and partridges. The plants obtained in his wanderings were the same, it is said, as those found on the more elevated parts of Socotra. Dragon's-blood, frankincense, and aloes were in abundance.

Mr. Bent [3] has quite recently ascended the hills behind Dhofár, at two places, accompanied by Mrs. Bent. He characterizes the view from the summit of the range as very curious. On the

---

[1] "The Ruins of El Balad," Journ. Roy. Geogr. Soc. xvi. 1846, pp. 187–199.

[2] Capt. Haines says the horns of the ibex had a curve of 3 ft.: a large head, doubtless of the same species, *Capra nubiana*, F. Cuv., in my possession, killed in the desert to the east of Heluan, near Cairo, measures 37¾ in. along the curve anteriorly. By the ounce, probably a leopard was meant.

[3] Geogr. Journ. vi. 1895, pp. 109–133.

side towards the sea the mountains are cut up by several deep gorges full of vegetation, and all the hills around, up to their summits, are covered with grass and clusters of trees, with here and there isolated groups of fig trees, their thick foliage being full of birds. He describes the aspect of the country in similar terms to Mr. Smith, designating it park-like, and mentions the presence of numerous herds of camels, goats, and oxen grazing over its pasture. He found the Garas living in caves on the hillsides. From the summit of the range, Mr. Bent saw the mountains sloping down towards the north and gradually becoming more and more arid until they merged in the yellow desert, which stretched as far as the eye could see, ending in the horizon in a straight blue line, as if it were a sea.

Rás Marbat, which forms the eastern limit of the plain of Dhofár, has at its base a granite plain four miles square and about 30 feet above the sea-level, with a group of low granite hills immediately below the headland itself, which consists of sandstone and limestone in the form of a precipitous tableland, 3400 ft. high, ascended by Carter, who has described its physical characters and geology.

Between the headlands Rás Marbat and Rás Nus there is a plain of dark igneous rock backed by an enormous cliff 3000 to 4000 ft. high, the seaward scarp of the tableland of the Subhán range of mountains. It descends in one step to the plain; but, when the granite headland of Rás Nus, 1200 ft. high, is rounded, the range is continued more or less to the north as a serrated ridge of at least four great peaks known as the Jebel Habareed, one of the most remarkable mountain-masses of this coast. Beyond this, to the east, the land suddenly sinks from 4000 ft. to 800 ft. in elevation, marking the termination, in this direction, of the wooded mountains, and of the fertile and populous region to the west, rich in flocks of goats, sheep, and camels, and in frankincense trees.

Rás Shirbetát, about 800 feet high, closes in the eastern side of the Bay of Khurya Murya. Here the coast is extremely desolate and almost devoid of vegetation, with the exception of a few date-palms, and brushwood in the ravines and dry watercourses giving cover to antelopes and hares. The largest ravine in the tableland of this bay is known as Wádí Rekót. It is also said to lead into the Hadramut, and, as far as it was examined by the officers of the 'Palinurus,' it appeared to be thickly wooded and well watered. The huge masses of rock in its dry watercourse fully attested to the strength of the current precipitated down it after heavy rain. A spring and a lake occur at its mouth, and on the latter widgeon and other wild ducks were shot by the officers of the Survey. The country of Jezzar, 120 miles inland, was described by the Arabs as abounding in the necessaries of life, and as yielding rich pasture for their flocks.

There are a number of islands, in the Bay of Khurya Murya, which were ceded to Great Britain by the Imaum of Maskat. One of them, known as Jébeliyah, has been described by Dr. Hulton [1] as perfectly barren, but the resort of sea-birds, and particularly of a gannet which, when he and his companions first landed, seemed inclined to dispute the ground with them. Lieut. Whish [2], writing about twenty years later, also calls the bird a gannet and states that it was extremely numerous and very noisy. It lays two eggs of a light blue tint upon the bare ground, merely clearing away the larger stones and collecting together a quantity of small gravel. The obstinacy with which the gannets defended their nests made them an easy prey. In consequence of their presence, the island was covered with large deposits of guano, which were estimated, in 1858, at 200,000 tons [3]. Wild cats were said to be seen sometimes on the rocks, and rats existed in great hordes, supposed to have been introduced by the wreck of some native vessel, as they were exactly like the common rat. Harmless snakes, described as whip-snakes, scorpions and centipeds were common [4].

From Rás Therrar, in Khurya Murya Bay, to Rás Jezirah, 170 miles to the east, the land subsides from 800 to 480 ft., but retains, generally, the appearance of a tableland, broken up however at Rás Shuamiyah by outbursts of igneous rocks. The whole of this part of the coast-line, with the exception of the sandy bay immediately to the west of Cape Jezirah, consists of a light-brown, barren, arid cliff of limestone rock, without a tree or even a mound to vary its outline; but, opposite to the small

---

[1] Journ. Roy. Geogr. Soc. xi. 1841, pp. 156-164 and map.
[2] Trans. Bombay Geogr. Soc. xv. 1860, pp. xxxvii to xl, with two plates.
[3] Buist (G.), Proc. Roy. Geogr. Soc. iv. 1859-60, pp. 50-57. In 1858 this island was leased by Government, for its guano, to a Liverpool firm of merchants.
[4] Buist (G.), loc. cit.

island, Hammar el Nafur, the coast presents a range of small
dark peaks rising gradually from the beach, probably the tops
of low igneous rocks. This island, 320 ft. high, is covered by a
multitude of shags.

From Rás Jezirah to the Bay Ghobat Hashish, opposite the
western end of the island of Masira, a distance of about
100 miles, the land gradually sinks to the level of the sea.

From Rás Abu Ashrin to Rás el Had, the most eastern headland of Arabia, the land rises somewhat, but is seldom more
than 100 ft. above sea-level. Along this extensive tract, which
is known to the Arabs as El Baetan, all mountains to the
west are lost sight of, but, in places, it rises into rounded, white
sand-hills, 200 ft. in height, among which may be observed dark
isolated peaks of similar elevation, whilst, in other parts, it is
simply a plain covered with salt efflorescence. This low desolate
tract is the eastward prolongation of the great sand desert of
Central Arabia.

The low land between Ghobat Hashish and Rás Abu Ashrin
is destitute of vegetation beyond some scattered tamarisks,
salsola bushes, and a few tufts of grass, but is sufficiently green,
to the eye of an Arab, to entitle it to the name it bears.

As the island of Masira lying off this bay is the only locality
on the south-east coast of Arabia, besides Makallah, that has
appeared in zoological literature, a few facts connected with it
may be of interest. It is situated about 100 miles to the west
of Rás el Had, and is 38½ miles in length and about 9 miles in
breadth, at its widest part. A range of mountains 600 ft. high
traverses it longitudinally and sends out spurs to the principal
capes, while shorter ridges branch out all over the island, more or
less rocky and irregularly pointed. With the exception of a few
dwarf babul and tamarisk trees, and matted grass in level places,
and a trace of small herbs in the mountains, it is essentially
barren ; but in the centre of the island there are a few date-palms,
as it is partially peopled. The miserable inhabitants own some
sheep and goats, and the usual domestic animals, the dog and cat.
The only wild animals known to Carter were a gazelle, and a
rabbit, half the size of the wild rabbit of Europe. Reptiles also
were present, but only one species is known, namely, the little
rock-gecko described by Dr. J. E. Gray as *Spatalura carteri* =
Pristurus carteri. Between the island and the mainland there
is a channel about ten miles wide, very shallow, and with several

islets lying in it. On the banks, around one of these islets surrounded with mangroves, myriads of wading birds such as
flamingoes, curlews, plovers, &c., congregate to feed at low
water. The island is strewn with the bones of turtle, as the
inhabitants largely use that animal as food.

Two remarkable mountains called Jebel Saffan lie within a
mile and a half of the shore at Rás el Had, with some hillocks
around them. They are the only mountains at the extreme
eastern point of Arabia, which is otherwise flat. There are two
khores leading into basins of considerable size, the southern and
eastern shores of the larger being low and swampy, and over-
grown with mangroves.

This completes a rapid sketch of the coast of Southern Arabia
from Aden to Rás el Had, but after the latter point is rounded
the following are the broad features of the coast-line northwards.
To the west a range of mountains rises from the plain in two
spurs, one 2700 ft. high and close to the coast, and the other the
Jebel Jallan, about 20 miles inland, and 3800 ft. in height. As
they run north they shortly unite and continue parallel to the
coast, with an elevation of about 4000 ft., and are precipitous
towards the sea, from which they are distant nearly eight miles.
About 70 miles north of this, the range is suddenly interrupted by
a narrow gorge known as the Devil's Gap, which is the opening
of a great valley called Makallah Obar, that runs up to the mountains of Oman. The range on the north of the gap rises suddenly
to 6228 feet above the sea, and trends to the north-west, with a
maritime plain between it and the shore; but within fourteen
miles of Maskat the shore-land becomes a confused mass of hills
and ridges with escarped precipices. To the west of Maskat the
main range is 40 miles inland, and 6000 ft. high. It is prolonged,
under the name of Jebel Akhdar, to Cape Massendam, at the
southern entrance to the Persian Gulf.

Maskat has become well known as a locality for reptiles,
through the energetic labours of Dr. Jayakar. It is rich in
reptilian life, but probably not more so than the area between
Makallah and the Hadramut.

The diversity of the physical characters of South-East Arabia,—
as seen in its generally barren maritime plain, varied occasionally, however, by the presence of tamarisks, acacias, and palms;
its cultivated and watered valleys running to the south from the
sterile mountain plateau, with nooks of sparse vegetation at

their heads; its deep and great cañons trending to the north, covered here and there with groves of palms and zizyphus, and richly cultivated fields; its nearly sand-choked valleys from the great desert; the fertile plain of Dhofár with its streams and lakes, its wooded uplands, and its grassy and park-like higher slopes,—offers conditions favourable to reptilian life, of which we now gain some insight, thanks to Mr. Bent's Expedition into the Hadramut.

Wellsted, who resided some weeks at Makallah, in 1834, says that the term Hadramut is a corruption by Europeans of an Arabic word meaning sudden death, and describes the region as " an extensive valley about 60 miles in length running nearly parallel to the coast." Mr. Bent, the most recent traveller in this part of Arabia, defines the Hadramut in almost similar terms, saying it is " a broad valley running for 100 miles or more parallel to the coast," and that " in the language of the Himyars it meant the enclosure or valley of death."

The Himyaritic inscriptions discovered by the officers of the 'Palinurus' at Hisn Ghoráb and Nakab el Hajar drew the attention of philologists to this part of Arabia, and led Baron Adolph Wrede to make his eventful journey, of 1843, in search of further material for the elucidation of the linguistic and historical problems that had been raised by the decipherment of these inscriptions. Similar reasons also induced Herr Leo Hirsch to enter the Hadramut, in July 1893, and Mr. and Mrs. Bent, in the latter part of the same year.

In the descriptions of the wanderings of Wrede and Hirsch we look in vain for any information bearing on the fauna of the region they visited, which is also unfortunately true of the writings of the officers of the ' Palinurus ' with the exception of the mention, at rare intervals, and in the most general terms, of antelopes, hyænas, hares, cats, and rats, and, in equally vague terms, of some birds.

Mr. and Mrs. Bent, however, started accompanied by a qualified botanical collector, Mr. Lunt, from the Kew Gardens; and by an Arab zoological collector provided by myself, and to whom I had given full instructions regarding the importance of keeping an accurate record of the locality in which each specimen was collected; but unfortunately he failed to attend to this, and I am therefore not in a position, except in one or two cases, to say more than that the specimens were collected between Makallah and the Hadramut Valley, and between that and the coast as far east as Shehr.

The accounts given by Wrede, Hirsch, and Mr. Bent of the features of this portion of Arabia may be summarized in a few words, after the general description I have given of the coast-line derived from the labours of the Officers of the Iudian Navy, now, I am sorry to say, almost forgotten.

The maritime plain or *Sahil* is narrow at Makallah, and the mountains rise abruptly from it, traversed on their seaward aspect by short and steep ravines and valleys.

Hirsch has given a graphic description of the route generally followed by caravans passing from Makallah to the Hadramut, and over which he travelled. Mr. Bent followed practically the same route; and Wrede, in 1844, ascended from Makallah to the plateau, by the same line of country, to reach Khoreba, on the west side of the Wádí Doáu, which he made his headquarters.

This route lies along the shore for a short distance and crosses a depression into which the sea at times penetrates aud into which a number of small *wádis* open. Further on, it passes the village of Bagrin, situated on the borders of a fertile *wádi*, the sides of which are clothed with an exuberaut growth of plants, richly watered by streams that trickle down the mountain-sides, and are carefully diverted for irrigation purposes. The Wádí Sidéd is afterwards followed, opening and contracting at places, but hemmed in on every side by high and dark mountains. A number of villages are passed iu this part of the route, and as it progresses the road rises more and more, overtopped to the left by mountaius, but it afterwards lies between high parallel ranges. Still further onwards the mountains of Ghail ba Wazir are passed on the right, with great precipices and rocky abysses, aud, beyond this, the Wádí Howari is entered, a long valley running up to the west and north. It is defined on the left by a high range of mouutains rising to 2000 feet above it, and in places assuming the appearance of gigautic castles erected by mau. Higher up, it becomes strewn with huge isolated masses of rock fallen from the mountains overshadowing it, aud as it is further ascended the grandeur of the scene increases, the cliffs on all sides rising perpendicularly and the mountains projecting majestically. This mountainous district is knowu as the Ghail Halka, and on the right of the valley lies the village of that name, surrounded by cultivation rising in terraces on the mountain-side

and watered by streams diverted into channels of irrigation. Still
ascending amidst these magnificent surroundings, the traveller
at last emerges on a vast plateau over 4000 feet high, great level
tracts of which are destitute of even a blade of grass and thickly
covered by small black stones, while throughout its extent it is
studded over, more or less, with low isolated hillocks, forming a
monotonous, dreary expanse, the horizon unbroken by a single
mountaiu-top. In traversing this plateau, it is found to be cut
into by numerous *wádís* running towards the north, and in their
beginnings mimosa, frankincense, and myrrh shrubs are found,
with other scanty vegetation, and in these localities an occasional
Bedouin woman may be met with tending her hardy but half-
starved flock of goats. Three days are spent crossing this
featureless, gloomy, untenanted desert towards the valley down
which the route lies to the Hadramut. This plateau is essentially
waterless, no stream or spring being present in any part of it,
but as occasional storms burst over it, tanks exist along the
route for the storage of the water; but, owing to the rapid
evaporation in this dry climate, these reservoirs are usually found
to be empty, except immediately after rain. When the traveller
reaches the margin of the plateau, where the route descends into
the Wádí Doán, an astonishing and unlooked-for scene opens out
before him, not distinctive of this valley alone, but common to
nearly all the many long valleys that pursue a northerly course
to the great Valley of the Hadramut, that is, to the Wádí Masílah.
Standing in such a spot, the plateau is found to dip down perpendicularly for 1000 to 1500 feet into the valley below, and in some
parts the cliffs stand out like a succession of gigantic castles;
but they generally terminate below in a slope of disintegration
on which the towns and villages are built, the bottom of the
valley being cultivated and covered with extensive groves of date-
palms. Wrede describes a flowing stream in that part of the
Wádí Doán where he entered it, 20 feet broad, enclosed by high
walled embankments and winding through fields laid out in
terraces; and Hirsch, who descended into the valley at Sif, says
that the channel of the river, when viewed from above, stretched
like a white thread through the valley, and into it he saw flowing
the Al Aisar from the south; the soil carefully divided out and
cultivated, with plantations of palms, and *Zizyphus spina-christi*
everywhere.

Mr. Bent observes that "the first peep down from the edge of the plateau into these very highly cultivated gullies is most remarkable, quite like looking down into a new world after the arid coast-line and barren plateau." The water-courses in these valleys are generally dry, and if running water occurs in the upper parts of any of them it ultimately becomes lost in the sands, but after heavy rain the water from the plateau is precipitated into the valleys and over the cliffs defining them. Water is always to be found on the level flats of these valleys; but Mr. Bent states, that in the Valley of the Hadramut proper, into which these valleys open, water for drinking purposes and for cultivation is only to be obtained by sinking wells.

The great Valley of the Hadramut, in the neighbourhood of its capital, Shibam, opens out into a wide plain, valleys entering it from the west, north, and south, the main valley being continued eastward to the sea where it opens at the town of Sibût, doubtless receiving many tributary valleys in its course: its seaward opening being one of the grandest on the coast. The level portions of the northern valleys of the plateau, and of the Wádí Masílah itself, are more or less covered with sand, while those running down from the great sand desert of the interior are choked with it, and as they are traced to the north, Mr. Bent says, the sand increases and becomes shifty and loose in places, and the hills on either side diminish in height. Wrede has given a description of a most remarkable accumulation of loose sand on the margin of the desert near Sava. He reached it from Khoréba by the Wádí Amd and the town of Haura at the upper end of the Hadramut Valley, where he ascended the plateau for the second time, and then descended upon Sava in the Wádí Rakhiab. He says that the desert, a day's journey from Sava, "presents an astonishing sight, consisting as it does of an immense sandy plain that gives it the appearance of a moving sea. Not a trace of vegetation, be it ever so scanty, appears to animate the vast expanse—not a single bird to interrupt with its note the calm of death." This portion of the margin of the desert, according to Wrede, lies 1000 feet below the plateau.

Hirsch, and Mr. Bent returned to the coast by the Wádí Adim, which Mr. Bent says differs from all the other valleys of the Hadramut, running into the plateau from the north, in that it ascends the plateau gradually. It is watered by a mountain stream, is very fertile and full of palm groves.

I take this opportunity to express my great indebtedness to Mr. Bent for having permitted my collector to accompany him on his Expedition.[1]

---

[1] The following is a list of the published descriptions of the Invertebrates collected on the Expedition :—

1. " On the Insects other than Coleoptera obtained by Dr. Anderson's Collector during Mr. T. Bent's Expedition to the Hadramaut, South Arabia." By W. F. Kirby, F.L.S., F.E.S.—Journ. Linn. Soc., Zool. vol. xxv. 1895, pp. 279–285.
2. " On the Coleoptera obtained &c." By C. J. Gahan, M.A.—Journ. Linn. Soc., Zool. vol. xxv. 1895, pp. 285–291.
3. " On the Arachnida and Myriopoda obtained &c. ; with a Supplement upon the Scorpions obtained by Dr. Anderson in Egypt and the Eastern Soudan." By R. I. Pocock.—Journ. Linn. Soc., Zool. vol. xxv. 1895, pp. 292–316, pl. ix.

PART II.

REPTILIA AND BATRACHIA

COLLECTED ON MR. J. T. BENT'S EXPEDITION

TO THE

HADRAMUT.

REPTILIA.

LACERTILIA.

GECKONIDÆ.

STENODACTYLUS (CERAMODACTYLUS) PULCHER, n. sp.

1 specimen.

Body somewhat slender; head rather elongately oval; snout pointed; eye large; ear an oval slit, directed obliquely. Nostril slightly tumid, formed by the rostral, first labial, and three nasals. Eleven upper and ten lower labials; mental large, rounded posteriorly, and projected backwards beyond the first lower labials. Limbs moderately long and slender; fingers not long, rather broad; toes moderately long, not narrow. The fore limb when laid forwards reaches the snout, and, when stretched backwards, falls short of the groin; the hind limb reaches somewhat beyond the axilla. Under surface of digits covered with very minute scales, feebly imbricate, and obscurely dentate anteriorly, and arranged in oblique rows of 8 scales to a row; a few well-defined transverse lamellæ towards the tips of the digits, where the scales are less numerous; upper surface of the digits covered with seven rows of smooth, feebly imbricate scales, the outer row modified on the fingers so as to form a feeble fringed edge, much more marked on the toes, especially on their outsides. Tail cylindrical, not thick, gradually tapered to a not very fine point, shorter than the body and head. Body covered with minute, rounded, slightly convex, juxtaposed scales, very obscurely granular, larger on the sides than on the middle of the back,

smallest on the occiput, the scales on the snout about the size
of those on the sides, or a little larger, and more markedly
granular than any of the other scales. Scales on the limbs
slightly larger than those on the body; scales on the tail arranged
in rings, larger than the body-scales, smooth, or minutely keeled;
scales on the under surface of the head minute, rounded granules;
those on the under surface of the trunk about the size of the
dorsal scales, somewhat oval, juxtaposed, and more or less granular. No præanal pores in the females.

General colour pale fawn, rather reticulately spotted with dark
brown on the head, and with three interrupted, broken, narrow,
brown lines on the back, and a narrow, rather feeble pale brown
line from behind the eye along the sides; the upper labials with
brown centres, and with the scales on the snout minutely speckled
with brown; a few dark spots on the thighs, and the upper surface
of the tail barred with the same dark colour, a round white
spot, as in *Stenodactylus elegans*, Fitz., alternating with the bars.
Under surface pure white.

Measurements.

Snout to vent .... 30·5 millim.[1]
Tail .... 26 „
Length of head .... 10 „
Width of head .... 9 „

This species differs from *S.* (*C.*) *doriæ*, Blanf, in its more
depressed body, more numerous scales on the under surface of
the digits, more tumid nostril, more elongate head, and more
pointed snout.

Although *Ceramodactylus doriæ*, Blanford, has five rows of
small imbricate scales on the under surface of the third toe, these
scales as they approach the tip tend to arrange themselves, and
do arrange themselves, in the same way as in *Stenodactylus
elegans*, Fitzinger, that is to say, the gradual passage of the
central rows of scales into transverse lamellæ is distinct and
present, so that the distal end of the digit of *Ceramodactylus* has
the structure distinctive of the entire digit of *S. elegans* Fitz.
In *Ceramodactylus affinis*, Murray, the scales on the under surface
of the digits are not so well marked off, into central lamellæ and
lateral scales, as they are in *S. elegans*, but in this intermediate
character serve to connect *C. doriæ* with the latter; and as there

[1] All measurements throughout this paper are in millimètres.

are no other characters separating them generically, there does not appear to be any reason why *Ceramodactylus* should retain more than subgeneric rank.

STENODACTYLUS (CERAMODACTYLUS) DORIÆ, Blanford.

Ceramodactylus doriæ, Blanford, Ann. & Mag. N. H. (4 ser.) xiii. 1874, p. 454; East. Persia, vol. ii. Zool. & Geol. (1876), p. 353, pl. xxiii. fig. 2: Blgr. Cat. Liz. Brit. Mus. i. 1885, p. 13, pl. ii. fig. 4*.

Two specimens agreeing with the types.

BUNOPUS BLANFORDII, Strauch.

Bunopus blanfordii, Strauch, Mém. Acad. Imp. St. Pétersb. (vii. ser.) xxxv. no. 2, 1887, p. 61, pl. figs. 13 & 14.

8 ♂ and 7 ♀.

This species has hitherto been recorded only from Egypt. Two specimens were obtained by Erber and described by Strauch, and are preserved in the Museum at St. Petersburg. I am indebted to Prof. Pleske, through the kind assistance of Mr. Boulenger, for the opportunity I have had of comparing one of the types with these specimens from the Hadramut. There can be no doubt regarding the specific identity of the African and South-East Arabian specimens.

This gecko is of considerable interest, as it is the only species that illustrates the passage of præanal into femoral pores. A line of enlarged scales stretches across the præanal region and is prolonged on to the thighs, in the position occupied by the femoral pores of other lizards. In the accompanying table, I have given the total number of pores. In the cases of the low numbers, the pores are essentially præanal, but, in those in which the numbers are higher, the pores pass on to the thighs, and, in the very highest numbers, may be seen in interrupted series extending nearly to the knee. This interrupted character and their extension over varying lengths of the thighs are of considerable interest.

Bunopus tuberculatus, Blanford, and *B. blanfordii*, Strauch, have both six rows of scales round the middle of the third toe, viz., five rows of scales all of which may be referred to the dorsal series, although one is lateral in position, and a longitudinal median row of lamellæ on its under surface. In the former species, the lamellæ are somewhat swollen and tubercular, whereas, in the latter, this character is but little marked, but the free

Bunopus blanfordii, Strauch.

Sex.	Snout to vent.	Tail.	Length of head.	Width of head.	Length of fore limb from head of humerus.	Length of hind limb posteriorly.	Praeanal and femoral pores.	Locality.
♂	49	63	15	9·5	20	26	20	Hadramut.
♂	49	.....	14·3	9	19	26·5	31	,,
♂	47	63	15	9·5	20	26	20	,,
♂	43	.....	13	9	18	24·4	17	,,
♂	42·5	54	12·6	8·4	16·5	22·6	18	,,
♂	41	48·5	12·8	8	16·9	22·4	14	,,
♀	40·5	.....	12·3	7	17·3	22·5	0	,,
♀	40·3	53	13	8	17	21·8	0	,,
♂	39	.....	12	7	16·8	22	10	,,
♀	36	.....	11·4	7	16	22·2	0	,,
♂	36	46	11·5	7·3	15·4	20·7	15	,,
♀	34	.....	10·2	7	15	19·5	0	,,
♀	33·5	.....	10·9	7	14·8	18·7	0	,,
♀	31·5	.....	11	7	13·8	17·3	0	,,
♀	30	39·8	10	6·5	13·3	21	0	,,

borders of the plates project and show evidences, under a hand-lens, of tridentation and swelling. The scales of the side of the digit in no way differ from the other dorsal scales, and conse-quently there is no true denticulation of the digits, but, of course, when seen in profile, the lateral scales project the one over the other. *Bunopus* has thus a simpler form of digit than *Steno-dactylus*; and as other differences manifest themselves in the form and scaly covering of the body, and in the shape of the tail, in both of which respects it resembles *Gymnodactylus* rather than *Stenodactylus*, it would seem to merit generic rank between these two genera, as held by Blanford and supported by Strauch.

PRISTURUS RUPESTRIS, Blanford.

Pristurus rupestris, Blanford, Ann. & Mag. Nat. Hist. (4) xiii. 1874, p. 454; East. Persia, vol. ii. Zool. & Geol. (1876), p. 350, pl. xxiii. figs. 1, 1 a; Proc. Zool. Soc. Lond. 1881, p. 465 : Murray, Vert. Zool. Sind, 1884, p. 365, pl. fig. 1 ; Boulenger, Cat. Liz. B. M. i. 1885, p. 53.

4 ♂ , 3 ♀ , and 2 juv.

Head rather short and moderately high ; snout variable, more pointed in some (Socotran examples) than in others (Maskat, and Hadramut Expedition), exceeding the interval between the posterior border of the eye and hinder margin of the ear-opening, and equalling the posterior orbital interspace, or nearly so ; fore-head flat, not concave ; eye moderately large ; nostril defined by the rostral and two or three nasals, the uppermost the largest ; rostral large, cleft above, twice as broad as high ; seven or eight upper labials; mental large, triangular, and broader than the rostral; five to six lower labials; no chin-shields, but a few scattered enlarged granules behind the mental and labials; ear situated below the level of the gape, small, oval in outline, and placed obliquely. Limbs long ; the fore limb reaches the end of the snout, and, when laid backwards, touches the groin, or falls short of it ; the hind limb reaches the ear. Tail laterally com-pressed, longer than the body and head, with a low dorsal crest of flat spines not extending on to the back, the mesial line of the under surface having no crest, but a line of enlarged projecting scales. In the female, the dorsal crest is very rudimentary. Body covered with minute granules, largest on the upper surface of the snout, especially in Socotran specimens. Scales on the sides of the tail larger than the body-granules, and arranged more or less in verticils; scales on the chin and throat minute, as small as the body-granules; those of the belly larger than the body-granules, but smaller than the scales on the upper surface of the snout.

Colour olive-grey; the back and sides with rufous spots, forming interrupted longitudinal lines, those on the back larger than those on the sides and with a white hinder margin [1]. A pale or light reddish band down the centre of the back. Sometimes a dark band from the nostril to the eye, and prolonged along the temporal region. The sides generally black-spotted, and the throat more or less marked with transverse, somewhat wavy, black and white bars.

This species is closely allied to *P. flavipunctatus*, Rüppell, but is distinguished from it by its generally longer hind limbs, and by the large and polygonal convex scales covering the snout.

The lizards from Socotra which have been referred to this species have a much more pointed and considerably longer snout than the types, and from the pronounced character of this variation, they would seem to be entitled to rank as a subspecies. The typical form of *P. rupestris*, Blanford, has hitherto been recorded only from Kharij Island, in the Persian Gulf, near Bushire, and from Maskat.

The larger ♂ measures as follows:—

Snout to vent	32	millim.
Vent to tip of tail	53	„
Length of head	9	„
Width of head	5·5	„
Length of hind limb	21	„

? PRISTURUS COLLARIS, Steindachner.

Spatalura collaris, Steindachner, Novara, Rept. 1867, p. 20.
Pristurus collaris, Blgr. Cat. Liz. Brit. Mus. i. 1885, p. 55.
39 specimens.

Head short and high; forehead flat, or convex antero-posteriorly; snout short, but longer than the distance between the eye and the ear, sharply pointed, beak-like; nostril perforated in a single, prominent, rather swollen, crescentic shield, the horns of the crescent either meeting behind the opening, or separated by one or more head-granules; occasionally, the nostril

---

[1] Blanford's description of fresh specimens.

is defined by two nasals, an upper and a lower, separated from each other posteriorly by head-granules ; ear-opening very small, obscure, oblique in position, the lower border being anterior ; rostral broad, pointed, convex from before backwards, in the form of a beak, and with a well-defined groove in the mesial, dorsal line of its proximal half. Generally 7 upper and 7 lower labials, but there may be as many as 8, and as few as 6. Limbs moderately long ; the hind limb when laid forwards reaches to the front border of the eye. Toes moderately long. The upper surface covered with minute, slightly convex granules, somewhat larger on the front of the head, but not markedly so, and smallest on the nape of the neck. The granules on the front of the fore limb and thigh are somewhat large and imbricate. The scales on the under surface of the body are larger than those on the back, and there are no erect spiny scales on the median line of the belly. Tail laterally compressed, not tapering to a fine point, but either truncated with a rounded end, or abruptly pointed, and covered with subquadrangular, flattened scales, larger than those of the body, and arranged, more or less, in verticils, with a rather feeble, serrated ridge along the dorsal and ventral lines.

General colour (in alcohol) pale greyish fawn, grey-brown, or even grey. Six or more quadrangular, transverse, brownish markings along the back from the nape to the sacral region, sometimes with pale posterior margins, and occasionally divided down the back by a pale mesial band. Externally to these dark squares, there are from 6 to 7 parallel lines of red spots, either rounded, or linear, the upper lines beginning behind the eye and the lower ones in the axilla. The labials are generally more or less blurred with blackish, this colour also invading the sides of the head, with yellowish granules intermixed. A narrow, purplish-black collar from side to side across the neck. Upper surfaces of the limbs more or less barred with black. Middle of the throat, chest, belly, and under surfaces of the limbs whitish. The tail barred like the back.

Measurements of an adult.

Snout to vent	52	millim.
Vent to tip of tail	48	,,
Length of head	15	,,
Length of hind limb	40	,,

Dr. F. Werner, of Vienna, has been so good as to compare two of the foregoing specimens with the types of *Spatalura collaris*, Steindachner, preserved in the Vienna Museum, and with which, he informs me, they are perfectly identical. The specimens I sent to Dr. Werner had their tails entire and unrenewed, whereas the only one of the four types examined by him that possessed a tail had it reproduced. A rough sketch of this tail, with which I have been favoured by Dr. Werner, represents a tail of the same type as that of *Spatalura*, Gray. It is unquestionably a reproduced tail, crested above and below. The tail, however, of this species, when renewed for the second time, becomes nearly cylindrical and the crests disappear.

The types of *S. collaris* were described by Dr. Steindachner as having a dorsal crest on the body, but Dr. Werner, having informed me that my specimens, which have no trace of such a crest, are perfectly identical with the foregoing types, are we therefore to conclude that an error has crept into the description of the species?

The only particulars in which *P. carteri*, Gray, differs from *P. collaris*, Steindachner, are that it has a mesial, ventral patch of spiny scales, and that no collar is present. Dr. Steindachner did not know whence his specimens of *P. collaris* were obtained; whereas the types of *P. carteri* were from the Island of Masira.

HEMIDACTYLUS TURCICUS, Linn.

Lacerta turcica, Linn. Syst. Nat. 12 ed., i. 1766, p. 362.

2 ♂, 4 ♀, and 3 juv.

These specimens are all very pale-coloured, with one exception in which the dark brown markings of the body and the brown bands on the tail are very pronounced. Doubtless, if the physical appearances of the localities in which these specimens were obtained had been recorded, the light-coloured individuals would have been found either to have come from the pale *sahil* or from the nearly white limestone cliffs, and the darker specimen from dark-coloured rocks, as all geckoes are very adaptive in their colouring.

HEMIDACTYLUS FLAVIVIRIDIS, Rüppell[1].

Hemidactylus flaviviridis, Rüppell, Anderson, Proc. Zool. Soc. 1895, p. 642.

1 ♂, Shehr on the sandy maritime plain to the east of Makallah.

[1] Type examined.

This species, which was first described by Rüppell from a specimen obtained at Massowah, was shortly afterwards described by Duméril and Bibron from Bengal as *H. coctæi.* Klunzinger, in 1878, again recorded it on the coast of the Red Sea at Koseir, and since then it has been observed at Aden and at Maskat, and has been found at Fao and Jask in Persia.

AGAMIDÆ.

AGAMA SINAITA, Heyden.

Agama sinaita, Heyden, Rüpp. Atlas N. Afr. 1827, p. 10, pl. iii.; Dum. & Bibr. Erpét. Génl. iv. 1837, p. 509; A. Duméril, Cat. Rept. Mus. Paris, 1851, p. 103; Boettger, Bericht. Senck. Nat. Ges. 1879–80, p. 195; Blgr. Cat. Liz. Brit. Mus. i. 1885, p. 339; Boettger, Kat. Rept. Mus. Senck. 1893, p. 49.

Agama arenaria, Heyden, Rüpp. Atlas N. Afr. 1827, p. 12.

Podorrhoa (Pseudotrapelus) sinaita, Fitz. Syst. Rept. 1843, p. 81.

Trapelus sinaitus, Gray, Cat. Liz. Brit. Mus. 1845, p. 259; Günther, Proc. Zool. Soc. Lond. 1864, p. 489; Tristram, West. Palest. 1884, p. 154, pl. xvi. fig. 3.

Agama sinaitica, Rüppell, Mus. Senck. iii. 1845, p. 302; Bedriaga, Bull. Soc. Imp. Nat. Moscou, 1879, no. 3, p. 37.

Agama mutabilis, Blgr. (*non* Merrem), Cat. Liz. Brit. Mus. i. 1885, p. 338; Boettger (*non* Merrem), Kat. Rept. Mus. Senck. 1893, p. 48.

Agama sinaiticus, Hart, Fauna and Flora of Sinai &c., 1891, p. 210.

2 ♂, 3 ♀, and 1 juv.

Isidore Geoffroy St. Hilaire, at p. 128, and again at p. 136 of the 'Description de l'Égypte,' refers to Merrem's Tent. Syst. Amph., and states that Merrem's *Agama mutabilis* was founded on the lizard represented in the former work on plate 5. figs. 3 and 4, and that the term used by Merrem was a translation into Latin of the French name under which the lizard was figured. Merrem's work was published in 1820, so that plate 5 had appeared before, and Isidore Geoffroy's text after that year. The plate had been issued before 1817, as Cuvier refers to it in the first edition of the 'Règne Animal.'

The two figures of Merrem's *A. mutabilis* are characterized by having the fourth digit of each foot longer than the third, so it cannot possibly fall under that division of the genus with the occipital not enlarged, in which the third digit is longer than the fourth. On the other hand, the *Agama sinaita*, Heyden, has the third digit on both fore and hind limbs longer than the fourth.

It seems probable that plate 5. figs. 3 and 4, viz. Merrem's *A. mutabilis*, may be the lizard described by Reuss, from Upper Egypt, under the name of *A. inermis*, which, I believe, is the species Mr. Blanford[1] had in view as the one to which he also would restrict the use of Merrem's name *mutabilis*.

I have examined the types of *A. sinaita*, Heyden, and *A. arenaria*, Heyden, preserved in the Frankfort Museum. The type of the former is a male with no gular pouch, and with six large præanal pores, *i. e.* with the same number as occurs in the specimens in the British Museum referred to *A. arenaria*. The dorsal scales are small and imbricate, but feebly so, of very uniform size but very regularly decreasing in dimensions towards the sides, where they are very small, yet still feebly imbricate. They are quite smooth on the anterior part of the body, but the scales on the limbs and sacral region appear to have been keeled, but only feebly so. The features of this individual are the small size of the dorsal scales, the regularity of their arrangement, their little imbrication, and their generally hexagonal form. The scales on the ventral surface are almost as large as the central line of scales on the back, and are smooth, or feebly keeled, here and there. The scales on the outsides of the limbs are considerably larger than any of those on the body, are strongly imbricate, and markedly keeled. The limbs are long and slender, and the third digit of both limbs is the longest. The scales on the top of the head are large, juxtaposed, and smooth, and there is a spine at the posterior margin of the ear. The nostril is placed slightly above the canthus rostralis, and looks upwards and backwards. The naked ear is larger than the eye-opening. The tail is laterally compressed, and the scales are strongly keeled. The coloration is completely faded. *Habitat* : Arabia Petræa.

A. arenaria, Heyden, is represented in the Frankfort Museum by the two types from Upper Egypt presented by Rüppell. They do not appear to me to differ from *A. sinaita*, except in

---

[1] Eastern Persia, ii. Zool. & Geol. 1876, p. 316.

having their dorsal scales a little more strongly keeled. In both there are seven præanal pores.

The females from the Hadramut have distinct præanal pores, and also those from Suez and Heluan (see Table).

The specimens of this species from Lower Egypt also illustrate its variations. I have met with it on the plain of Suez and in the desert (Wádí Hoaf) at Heluan. The lizard from the former locality has its dorsal scales practically smooth, whereas those from the latter have distinctly keeled scales. The specimens in the British Museum referred to *A. arenaria*, Heyden, and also from Egypt, exactly resemble the lizard from the plain of Suez, whereas two specimens in the British Museum from Mount Sinai, and referred to *A. sinaita*, Heyden, correspond to one of my Heluan lizards, a young individual.

The adult female from Heluan (Wádí Hoaf) has the mesial line of dorsal scales very slightly, if at all, enlarged, considering the fact that in all specimens belonging to one or other of these varieties the scales gradually diminish in size towards the sides. In this specimen, however, the scales are decidedly imbricate and distinctly keeled. On the other hand, in the specimen from the plain of Suez the scales are only feebly imbricate, and carination is all but completely lost. The mesial dorsal scales hold almost the same proportions to the lateral scales as in the Wádí Hoaf female. In both of these specimens, and in the Hadramut examples as well, the ventrals do not vary in size; the slight difference between them is confined to the varying development of the dorsal scales. As a rule, the Hadramut specimens have the mesial dorsal scales decidedly larger than the ventrals, and all have distinctly keeled scales, but, among some of them, the difference in size between dorsals, laterals, and ventrals graduates in the same way as in the Wádí Hoaf female.

In view of these facts, and the exact similarity of these lizards in the other details of their external structure, the differences I have pointed out can only be regarded as illustrative of variation, but they present no stability to entitle them to varietal rank. It is only another example of the remarkable modifications to which the scales of many species of the genus *Agama* are subject, and which is perhaps most strongly pronounced in that strangely variable form *A. inermis*, Reuss, which in one of its phases could never be recognized under this specific term.

Agama sinaita, Heyden.

No. of specimen.	Sex.	Snout to vent.	Tail.	Length of head.	Width of head.	Length of fore limb from head of humerus.	Length of hind limb from behind.	Length of tibia.	Praeanal pores.	Locality.
518.........	♂	84	143¹	19	22	49	76	27	6	Hadramut.
38.........	♀	76	124¹	21	18·5	49	68	26	4	Plain of Suez.
515.........	♀	71	......	18	17	45	65	24	7	Hadramut.
519.........	♀	70	144	18	16	46	68	24·2	4	"
16.........	♀	65	140	18·5	17	44	65	24	4	Desert above Wâdi Hoaf, Heluan.
277.........	♂	62	147	17	17	41	65	24	4	"
520.........	♂	59	129	16	14	40	58	21	4	Hadramut.
517.........	♀	57	......	15	14	35	56	19·6	4	"
516.........	Juv.	46	96	12	11·5	28	41	15	...	"
14.........	Juv.	43	83	12	11·5	29	41	14·8	4	Desert above Wâdi Hoaf, Heluan.

¹ Tail not quite perfect.

In the two males from the Hadramut, there are three bright orange or reddish-brown bands on the back—the first on the nape of the neck, the second behind the shoulders, and the third on the loins. These bands are interrupted on the mesial line, and the first and last are narrow, but the second expands on the sides. Rusty-coloured bars occur at intervals on the tail. The heads are yellowish, and bluish.

From the list of Arabian reptiles appended to this paper, it will be seen that the species occurs at Maskat, Aden, the Hejaz, Akabah, and the Sinaitic Peninsula.

AGAMA FLAVIMACULATA, Rüppell.

Trapelus flavimaculatus, Rüppell, Neue Wirbelth. 1835, Rept. p. 12, pl. vi. fig. 1.

Agama agilis, Dum. & Bibr. in part, Erpét. Génl. t. iv. 1837, p. 496.

Agama leucostigma, Blgr. (*non* Reuss), Cat. Lizards Brit. Mus. i. 1885, p. 346; Boettger (part), Kat. Rept. Sammlung Mus. Senck. 1893, p. 49.

1 ♂ and 1 ♀.

AGAMA ADRAMITANA, n. sp.

3 ♂ and 2 juv.

Mr. Blanford, a good many years ago, described an Agamoid lizard from Abyssinia which he designated *A. annectens*, as he held that it served to connect *Agama* and *Stellio*, its tail conforming to that of the latter, whereas in other respects it agreed best with the former. A lizard closely allied to it occurs in the country between Makallah and the Hadramut Valley, but it presents certain characters which at once enable it to be distinguished from the Abyssinian species.

Head triangular; snout rather pointed. A prominent median ridge on the snout before the eyes in the adult male, less developed in young specimens. Head-scales of moderate size, and smooth. Nostril small, slightly below the canthus rostralis, directed outwards and backwards. Ear twice as large as the eye-opening. A prominent spiny eminence at the front border of the ear, and a few spiny scales above it; a spiny eminence at the lower border of the ear, and two at some distance behind its posterior border. A spine on the hinder aspect of the angle of the jaw, and a line of spiny scales along its outer surface, continuous with the lower labials. A strong group of spines on the post-temporal region. A short but strong, low, nuchal crest of about six spines, and a rosette of spines a little way external to its middle. Fifteen to seventeen upper and lower labials. Scales of the body small, imbricate, keeled, with sharp but short projecting points, the largest along the middle of the back, arranged more or less in transverse series; 119 scales encircling the middle of the body, and 58 rows occurring between the origin of the limbs. The scales on the limbs very much larger and more strongly imbricate and keeled than those of the body; the scales on the base of the tail nearly four times as large as the largest body-scales. The scales on the sides of the body are smaller than the ventrals, which generally have small sharp points, and are either feebly keeled or smooth. The scales on the tail are strongly keeled and terminate in short sharp points. On the base of the tail the scales are not arranged in segments, but, a short distance further back, the tail becomes segmented, each division containing about four annuli. The skin of the neck forms a loose longitudinal fold, (there is no true gular pouch), and is traversed transversely by a fold between the angles of the jaw, ending posteriorly in the true gular transverse fold. The upper surface and the sides of the neck are in loose folds. A fold arises from the rosette of spines external to the middle of the nuchal crest, and passes outwards and backwards a short way and terminates in a prominent spiny eminence, from which a fold crosses the upper surface of the neck to the corresponding eminence on the opposite side: in its course across, there are three rosettes—one external to the mesial line, another to its fellow of the other side of the neck, and the third on the mesial line immediately behind the nuchal crest. Another small fold arises at the prominent spiny eminence, and passes backwards to the front of the præhumeral pit, and ends in a few small spines, at which point it is joined by two small folds from the angle of the jaw. From the point of union of these folds, another passes upwards over the shoulder, along the side to near the sacral region, and in its course it is more or less beset with small spiny scales or rosettes. Immediately above the shoulder, a small spiny fold crosses up to the side of the neck. From behind the transverse nuchal fold, a series of small spines, set at intervals, extend as far back as the shoulder. The limbs are well developed, and the tibia is considerably longer than the skull. The wrist in all reaches in advance of the snout; but the tip of the fourth toe, in three, reaches the eye, whilst, in two, it is in advance of the eye. The digits are rather long, and laterally compressed. The fourth finger is only very slightly longer than the third, and the fourth toe than the third. The tail is somewhat slightly compressed beyond the base, but afterwards it is round and tapers to a fine point; it is about twice as long as the distance between the snout and the vent. Six to twelve præanal pores in the male, with callose scales on the abdomen.

Olive, mottled with brownish; blue about the eyes and along the labial line; the throat more or less reticulated with bluish lines. Underparts yellowish, but a few blue spots on the belly. A young male is olive, but the body has brownish markings, and bluish green on the head above, and bright blue below with dark blue lines; belly greenish yellow, mottled with blue; tail yellow at the base, olive distally, with about 18 brown bars. Another male is entirely blue above and brilliant blue below, the base of the tail yellow.

The rosette of spines on the post-temporal region, the short, but well-defined nuchal crest, with a rosette of spines on either side of it, near its middle, are all absent in *A. annectens*, Blanf., in which the lateral fold along the side over the shoulders is also practically absent, but, if feebly present, it never presents the small spinose rosettes that occur in the Arabian form. The scales also of the body are smaller in the Abyssinian species, in which there are 150 on the type around the middle of the body, and only 119 in the Arabian species.

A. adramitana is distinguished from *A. cyanogaster*, Rüppell, and from *A. nupta*, De Fil., by its much smaller scales.

PHRYNOCEPHALUS ARABICUS, Anderson.

Phrynocephalus arabicus, Anderson, Ann. & Mag. Nat. Hist. ser. 6, vol. xiv. Nov. 1894, p. 377.

1 ♂ and 1 ♀.

UROMASTIX (APOROSCELIS) BENTI, Anderson.

Aporoscelis benti, Anderson, Ann. & Mag. Nat. Hist. ser. 6, vol. xiv. Nov. 1894, p. 376.

3 ♂ and 3 ♀, Bagrin, 3 miles from Makallah.

As the absence of femoral pores is the only feature wherein *Aporoscelis* differs from *Uromastix*, there does not appear to be

d

any valid reason for its retention as a distinct genus. It may, however, be used to indicate the subdivision of *Uromastix* in which pores are absent, represented by the two species *U. princeps*, O'Sh., and *U. benti*, Anders. *Uromastix batilliferus*, Vaill., from its dentition and the form of its body, is unquestionably a member of the genus *Agama*, but with the tail of an *Uromastix*.

VARANIDÆ.

VARANUS GRISEUS, Daud.

1 ♀.

LACERTIDÆ.

ACANTHODACTYLUS BOSKIANUS, Daud.

16 ♂, 20 ♀, and 3 juv.

The scales round the bodies of these specimens vary from 35 to 48. The latter number leads into the type of fine lepidosis so characteristic of this species along the seaboard of Lower Egypt, where the scales range from 46 to 57. The accompanying table (pp. 35–37) contains the measurements of the largest specimens of this species yet recorded. I have tabulated the measurements of 199 specimens from a great number of widely separated localities, but not one attains to the dimensions of the largest Hadramut individual.

The coloration of some of them is somewhat different from that found in other localities, as the upper surface of a few of the adults is lineated with bluish-grey and reddish-fawn.

ACANTHODACTYLUS CANTORIS, Günther.

Acanthodactylus cantoris, Günther, Rept. Brit. Ind. 1864, p. 73 ; Anderson, Proc. Zool. Soc. 1895, p. 646.

5 ♂ and 9 ♀.

Snout elongate and acutely pointed, more so in some than in others. The frontonasals may form either a long or short suture, depending on the degree of elongation of the snout which also affects the length of the præfrontals ; four supraorbitals, the fourth generally consisting of one elongated piece with granules in front of it, but occasionally quite entire ; temporals elongate, and generally keeled ; anterior border of the ear with an outer row of enlarged scales, resembling truncated denticles, and an inner row as well, but the latter is occasionally feebly defined. The back, behind the shoulders, is covered with enlarged, imbricate,

Acanthodactylus boskianus, Daud.

Number of specimen.	Sex.	Snout to vent.	Tail.	Scales round body.	Ventrals.	Large scales between thighs.	Femoral pores.	Locality.
91	Juv.	33	65	40	10	10	17/17	Hadramut District.
149	♀	36	80	42	10	10	19/18	,, ,,
82	Juv.	36	75	38	10	10	20/20	,, ,,
163	Juv.	37	79	40	9	10	15/16	,, ,,
37 e ...	♀	38	80	35	8	10	20/21	,, ,,
122	♂	39	85	38	10	10	22/20	,, ,,
130	♂	40	88	43	11	10	21/20	,, ,,
96	♀	41	90	43	10	10	19/18	,, ,,
37 f ...	♂	42	......	43	10	10	16/16	,, ,,
131	♀	42	88	40	10	10	18/19	,, ,,
132	♂	43	96	40	10	10	19/20	,, ,,
23	♂	43	105	45	10	10	18/19	,, ,,

Acanthodactylus boskianus (continued).

Number of specimen.	Sex.	Snout to vent.	Tail.	Scales round body.	Ventrals.	Large scales between thighs.	Femoral pores.	Locality.
103	♂	44	100	37	10	8	21/21	Hadramut District.
121	♀	51	102	40	10	9	20/19	" "
38	♀	52	......	48	10	12	20/22	" "
120	♀	53	111	43	10	10	19/20	" "
37	♂	53	128	43	10	10	21/22	" "
102	♀	53	112	39	10	8	20/21	" "
156	♂	53	......	40	10	10	18/19	" "
136	♂	55	121	43	10	10	20/22	" "
155	♂	56	......	37	10	10	20/22	" "
244	♀	57	129	40	10	10	21/20	" "
185	♂	57	120	37	10	10	16/17	" "
92	♀	60	......	37	10	10	20/20	" "

„	„	„	„	„	„	„	„	„
„	„	„	„	„	„	„	„	„

Fraction	a	b	c	d	e	Sex	No.
18/18	9	10	42	.....	63	♀	118 .....
17/17	8	10	36	.....	71	♀	138 .....
18/16	10	10	38	.....	73	♀	137 .....
20/21	10	10	38	153	74	♀	117 .....
17/17	10	10	37	142	74	♀	111 .....
19	10	10	39	.....	76	♀	114 .....
21/20	10	10	36	.....	77	♀	113 .....
20/20	8	10	39	175	77	♂	37 a .....
18/16	10	10	35	152	78	♀	115 .....
17/16	10	10	35	153	80	♂	92 a .....
17/17	10	10	37	155	81	♀	110 .....
19/20	10	10-11	40	151	82	♀	109 .....
19/20	10	10	46	.....	82	♀	112 .....
19/18	10	10	36	180	82	♂	150 .....
17/19	10	10	36	190	86	♂	108 .....

Acanthodactylus cantoris, Günther.

No. of specimen.	Sex.	Snout to vent.	Tail.	Scales round body.	Ventrals.	Enlarged scales between thighs.	Length of third toe.	Femoral pores.	Locality.
276	♂	59	136	45	14	16	13	$\frac{21}{23}$	Hadramut.
293	♂	55	115	47	14	14	14·2	$\frac{21}{22}$	"
286	♂	52	127	45	14	13	14	$\frac{22}{22}$	"
37 e ...	♂	49	116	43	14	12	13	$\frac{22}{22}$	"
37 a ...	♂	49	......	43	14	12	13	$\frac{23}{22}$	"
37 b ...	♀	49	......	46	14	14	11·5	$\frac{20}{19}$	Gravid.
37 f ...	♀	47	......	47	14	14	11·2	$\frac{22}{21}$	"
37 g ...	♀	46	100	43	14	14	12	$\frac{20}{19}$	Gravid.
37 h ...	♀	45	......	43	14	12	11	$\frac{20}{21}$	Gravid.
37 d ...	♀	44	95	47	14	13	11	$\frac{19}{20}$	Gravid.
37 o ...	♀	40	89	43	14	12	11	$\frac{21}{19}$	"

Locality								Sex	Reg. No.
Hadramut.—Gravid.	19/19	10	11	14	44	.....	44	♀	37 *i* ...
"	19/19	10	11	14	42	90	42	♀	37 *j* ...
"	20/19	10·3	12	14	47	88	40	♀	92 *c* ...
Ramnuggur.—Types.	19/19	13	11	14	44	120	64	♂	60.3.19.1040.
"	22/20	13·7	13	14	45	140	63	♂	76.10.13.4 ...
"	21/15	15	12	14	41	153	69	♂	80.11.10.33...
Sind.	24/23	15	13	14	47	132	72	♂	26...
"	21/23	14	12	13	44	112	63	♂	28...
"	19/20	13	10	12	43	.....	54	♀	34...
"	19/19	12	13	12-13	44	113	59	♀	35...
"	19/20	12	12	12	47	117	59	♀	29...
Lineated, young.	22/21	11	12	14	49	104	48	♂	36...
"	22/22	13·3	10	12	39	92	46·5	♂	37 ..
"	20/20	9·8	13	12-13	41	72	46	♂	38...
"	19/20	8	11	12	38	65	32	♀	

Acanthodactylus cantoris (continued).

Number of specimen.	Sex.	Snout to vent.	Tail.	Scales round body.	Ventrals.	Enlarged scales between thighs.	Length of third toe.	Femoral pores.	Locality.
70.11.29.151.	♂	42	83	45	12	10	10·5	$\frac{20}{21}$	Hissar.—Lineated.
86.9.21.80 ...	♂	50	.....	48	14	12	11·3	$\frac{18}{19}$	Helmund.
86.9.21.77 ...	♂	74	142	45	12	13	14·5	$\frac{21}{18}$	,,
78 ...	♂	72	124	45	13	12	15·5	$\frac{20}{20}$	,,
79 ...	♀	63	.....	45	13	12	12	$\frac{18}{19}$	,, Gravid.
74.11.23.41...	♂	58	117	52	14	12	13	$\frac{24}{22}$	Dasht River, Baluchistan. Lineated, but not brilliantly.
42 ...	♂	51	102	57	13	14	11·6	$\frac{19}{18}$	,, ,, ,,
39 ...	♂	52	94	55	14	16	12	$\frac{21}{21}$	Mand, Baluchistan.
38 ...	♂	65		54	14	14	15	$\frac{18}{19}$	North of Bam, S.E. Persia.

Reg. No.	Sex							Locality
94.11.13.14 ...	♂	65	......	50	12	14	15	21/19	Jask, Persia.
15 ...	♀	54	102	54	13	11	11·8	17/19	Jask, Persia. cont, gravid. Lineated, adolescent "
16 ...	♀	52	......	54	14	12	11·5	17/18	"
95.5.23.47 ...	♂	54	87	40	14	12	13·5	19/19	Aden.
48 ...	♀	53	......	39	14	10	12	21/22	"
48 ...	♀	41	......	43	14	11	10·5	22/21	"
50 ...	♂ juv.	42	81	42	14	12	10·2	21/22	"
51 ...	♂	40	86	38	14	12	11	20	"
52 ...	Juv.	38	88	45	14	12	10	22/20	"
53 ...	Juv.	35	80	43	13	12	9·1	20/21	"
54 ...	Juv.	33	77	40	14	12	9	21	"
55 ...	Juv.	31	68	39	13	10	9	18/17	"
56 ...	Juv.	29	......	45	12	12	8	20/19	"
57 ...	Juv.	26	56	43	12	12	8	20	"

strongly keeled scales resembling those of *A. boskianus*, Daud.,
increasing in size to the root of the tail; generally 10 to 16 rows
of enlarged scales between the thighs; 38 to 57 scales round the
body; 12 to 14 ventral plates, broader than long, the higher
number being the most prevalent, but, as the ventrals pass very
gradually into the scales on the sides, it is sometimes difficult to
define the line of separation. An enlarged præanal surrounded by
large scales, but occasionally broken up. 17 to 24 femoral pores.
The fore limb is well developed, and reaches to the extremity of
the snout. The digits have an upper, an inferior and a lateral
plate to each of their sides; claws moderately long, compressed,
and sharply curved. The hind limb reaches to between the eye
and the ear, and even in advance of the eye, and its digits are long
and tapered, as in *A. scutellatus*, Aud.; the fringe is long as in
that species, and is longest ou the outer edge of the digits; the
hind claws are long, tapered, and little curved.

The coloration is much the same as in *A. boskianus*, Daud.
The adult may be uniform olive, greyish, or even brownish, sometimes sparsely covered with smaʌl black spots, tending to a
longitudinal arrangement in lines. The young is lineated with
eight black and seven white bands. The upper surfaces of the
limbs are generally covered with white spots on a dark ground,
and, in some semi-adults, there is a dark line along the back of
the thigh. Underparts white.

The typical form, from Sind and the Helmund, is considerably
larger than any of the specimens from South-Eastern Arabia.
A male from the former region measures 74 millim. from the
snout to the vent, whereas, from the latter locality, the largest
male is only 59 millim. Besides this difference in size, the typical
form has a somewhat longer, narrower, and more pointed head
and snout, but specimens are met with in which the head is not
so pointed, and in which the snout becomes obtusely rounded,
while in some examples from Sind the head is even still shorter,
but with a pointed snout. Specimens from Southern Persia
attain also to a greater size than those from Arabia, but fall
short of the dimensions of the Helmund lizards. The heads of
lizards of this species from Jask approach in their form more to
the Arabian than to the Sind specimens, the head being rather
short and broad, with a pointed snout, varying in the degree to
which the nasal portion is widened. There is also a marked
difference between the form of the head when the two extremes

are studied by themselves ; but when the series tabulated is considered in detail, the two are unmistakably linked together by intermediate variations.

The lepidosis throughout the area of distribution, with the exception of Baluchistan and South Persia, conforms generally to that of the typical form, but in these two regions the scales are somewhat smaller and consequently more numerous. In Sind the variation is from 38 to 49 ; in the Helmund 45 to 48 ; in Baluchistan and South Persia 50 to 57 ; in the Hadramut 42 to 47 ; and Aden 38 to 45. Were the specimens enumerated in the accompanying table (pp. 38–41) arranged according to the numerical sequence of their scales, they would form a practically unbroken series from 38 to 57.

The structure of its digits and its acutely pointed snout enable this species at once to be distinguished from *A. boskianus*, Daud.

EREMIAS GUTTULATA, Licht.

Eremias guttulata, Blgr. Cat. Liz. Brit. Mus. iii. 1887, p. 87 ; Anderson, Proc. Zool. Soc. 1895, p. 646.

1 ♀.

In this species the interparietal is almost always in contact with the occipital, but exceptions occur in which a well-developed small plate is interposed between them. It is rare, however, as I have only observed it in three cases, among 52 specimens from Egypt proper (Nile Valley) and the district of Suakin. One instance occurred at Luxor, and two at Durrur, north of Suakin. In the former locality I also met with two specimens in which two small scales existed side by side between the two shields in question. This specimen from the Hadramut has also a plate interposed between them. In its low number of ventrals, viz. eight, it resembles the variety described by Stoliczka as *E.* (*Mesalina*) *watsonana*, which had also a small plate interposed between the interparietal and occipital.

EREMIAS BREVIROSTRIS, Blanford.

Eremias watsonanus, Stoliczka, Proc. As. Soc. Beng. 1872, p. 125 (*nec ante*, p. 86).

Mesalina brevirostris, Blanford, Eastern Persia, ii. Zool. & Geol. 1876, p. 379.

Eremias brevirostris, Blgr. Cat. Liz. Brit. Mus. iii. 1887, p. 89.

14 ♂ and 4 ♀.

Head short, contracted before the eyes ; snout short ; nostrils

Eremias brevirostris, Blanford.

No. of specimen.	Sex.	Snout to vent.	Tail.	Scales round body.	Ventrals.	Longitudinal series of ventrals.
90...............	♂	41	99	35	10	32
151...............	♂	41	82	35	10	31
153...............	♂	42	...	34	10	30
94...............	♂	42	...	39	10	31
95...............	♂	41	83	34	10	31
97...............	♂	39	73	30	10	31
98...............	♂	36	73	31	10	31
124...............	♂	34	65	31	10	31
173...............	♂	40	...	37	10	31
79...............	♂	42	101	37	10	28
173 b	♂	34	70	37	10	32
73...............	♂	39	...	39	10	33
161...............	♂	35	...	39	10	31
78...............	♂	34	76	39	10	32
100...............	♀	39	70	31	10	32
160...............	♀	35	...	34	10	30
147...............	♀	40	67	42	10	34
173 a	♀	41	...	37	10	32
74. 11. 23. 82........	♂	36	51	43	12	30
80. 11. 10. 40........	♂	44	85	45	12	32

Eremias brevirostris, Blanford.

Infraoculars between labials.	Femoral pores.	
4·5	$\frac{14}{14}$	Hadramut.
5·6	$\frac{13}{13}$	,,
5·6	$\frac{12}{12}$	,,
5·6	$\frac{14}{14}$	,,
4·5	$\frac{14}{14}$	,,
{ 5·4 4·5	$\frac{13}{14}$	,,
4·5	$\frac{11}{11}$	,,
4·5	$\frac{13}{12}$	Hadramut. Occipital reduced to a granule and not wedged in between the parietals.
4·5	$\frac{13}{13}$	Hadramut.
4·5	$\frac{12}{12}$	Hadramut. Small plate behind the interparietal but not in contact with it, and widely separated from the occipital.
{ 4·5 5·4	$\frac{16}{14}$	Hadramut.
{ 4·5 5·4	$\frac{14}{14}$	Hadramut. A small plate behind the interparietal in contact with it, but not reaching the occipital, the latter wedged in between the parietals.
5·6	$\frac{12}{12}$	Hadramut.
{ 4·5 5·6	$\frac{13}{14}$	,,
4·5	$\frac{13}{11}$	,,
4·5	$\frac{12}{12}$	,, No occipital.
4·5	$\frac{12}{13}$	,,
4·5	$\frac{14}{14}$	,,
4·5	$\frac{13}{13}$	Tumb Island, Persian Gulf. Type of *C. brevirostris*.
{ 4·5 5·4	$\frac{14}{14}$	Kalabagh, Punjab. Type of *E. watsonanus*, Stol.

swollen, formed by three nasals (an upper, lower, and posterior), the latter small, and sometimes excluded from the nostril by the apposition of the other two at their inner points. Frontonasal grooved, considerably broader than long, shut off from the rostral by the nasals; two præfrontals, grooved in the mesial line; frontal deeply grooved longitudinally, rather narrow, its length equals the distance between its anterior border and the free margin of the rostral; anterior supraocular broken up; the second and third separated from the superciliaries by a line of granules; the fourth supraocular small, or broken up into small pieces with granules external to it; frontonasal pentagonal; interparietal quadrately oval, slightly smaller than a single frontoparietal; a small plate behind the interparietal, in contact with, or not in contact with, the interparietal and occipital, but sometimes wholly absent; the occipital is occasionally absent; two loreals, the first long and narrow; ocular disk more or less broken up; temporals granular, minute, and smooth; a line of elongated scales along the parietals. From 4 to 6 labials before the interocular; infraocular in the labial margin, but occasionally excluded by portions separated off from itself. Ear with an enlarged curved scale at its upper border. Body-scales granular, smooth, but more or less feebly keeled on the loins, as they approach the caudal scales, or they may be smooth throughout. 30 to 45 scales round the middle of the body; scales on the upper surface of the tail strongly keeled, those of the under surface smooth, or feebly, obtusely keeled; small scales, and large plates of the limbs, smooth. 10 or 12 rows of ventral plates, depending on the degree of development of the two outer rows; 28 to 33 in a longitudinal line from the collar to the femoral pores; the two median rows slightly broader than long. 11 to 15 femoral pores; an enlarged præanal with a semicircle of enlarged scales. Limbs and digits slender; the fore limb, when laid forwards, reaches to between the eye and the nostril, and the hind limb to between the shoulder and the ear. Tail variable, generally more than twice as long as the body and head, but sometimes shorter, tapered to a very fine point.

Fawn-coloured above, or pale yellowish, or greyish brown; a broadish brown lateral band from behind the eye, more or less pale-spotted, with indications of three longitudinal dorsal lines of small quadrangular, sometimes rather obscure brown spots, with intervening lines of smaller white spots, alongside of them; a more or less interrupted, narrow, white line along the upper margin of the dark lateral band, and a more or less orange line below it; a dusky line from the nostril to the eye, with the upper labials faintly speckled with brown; upper surface of head immaculate, or finely spotted with brown; the lateral brown band is prolonged on to the sides of the base of the tail; upper surface of limbs marbled with brown, with one or two whitish spots on the hind limbs. Under surface pure white.

These specimens have been compared with the types of the species, from which they differ in having the scales immediately external to the 10th row of ventrals partaking more of the character of dorsal than of ventral scales. Some of these scales, rarely however, assume the character of ventrals, so that eleven ventrals are present in some, the odd number being due to the scale of the opposite side not having taken on the full characters of a ventral. These cases of asymmetry are not recorded in the accompanying table, but their existence suggests that the difference between the types and these Hadramut specimens, in the number of their ventrals, will be bridged over when more materials from additional localities are examined.

Scincidæ.

Mabuia brevicollis, Wiegm.

Mabuia brevicollis, Anderson, Proc. Zool. Soc. 1895, p. 646.

2 ♂, 3 ♀, and 1 juv.

The two males are distinguished by the presence of pure white spots on the head and on the anterior part of the body, while the females are not. One female is of considerable interest, as the way in which the dark spots are arranged in obliquely disposed lines across the body, and the presence on some of the scales of a white central dart, recall the coloration of *Chalcides ocellatus*, Forskål. This type of coloration is also occasionally present in *Eumeces schneideri*, Daud., of which Mr. Boulenger has shown me some specimens, from Hoana, near Alexandretta, with the characteristic markings of *C. (G.) ocellatus*, Forskål, so perfectly reproduced that the lizards might, at first sight, be mistaken for it.

Mabuia brevicollis, Wiegm.

Sex.	Snout to vent.	Tail.	Length of head.	Width of head.	Length of fore limb from head of humerus.	Length of hind limb.	Scales round body.	Ear-lobules.	Relation of anterior loreal to labials.	Relation of postnasal to labials.	Position of nostril.	Axilla to groin.	Shoulder to anterior angle of eye.
♀ ...	135	...	26	19	39	51	32	R. 3 L. 4	2 & 3	over 1 & 2	over 1st labial	65	39
♂ ...	126	181	28	22	40	47	32	4	2 & 3	L. over 1 & 2 R. „ 1	„	61	37·8
♂ ...	120	220	27	20	39	48	34	R. 3 L. 4	2 & 3	1	„	59	36·1
♂ ...	106	...	25	17·3	38	44	32	3	2 & 3	1	„	54	28·2
♀ ...	97	168	21	17	31	38·3	34	4	2 & 3	1 & 2	„	50	27
♀ ...	82	132	17	13	26	33·5	34	4	2 & 3	1 & 2	„	39	23·9
Juv. .	56	...	13	8	19	24	34	R. 3 L. 2	1 & 2	1	„	26·5	16·2

Scincus conirostris, Blanford.

Scincus conirostris, Blanford, Proc. Zool. Soc. 1881, p. 677, fig. 1 ; Blgr. Cat. Liz. Brit. Mus. iii. (1887) p. 391.

3 specimens.

These fine specimens have been compared with the types, with which they agree. The ear is covered with two large fringed scales, but, at the same time, it is perfectly distinct. The supra-- nasal suture is much broader in some than in others, but it always effectually excludes the frontonasal from contact with the rostral. Two specimens have 26, and the other 28 rows of scales round the body.

Chalcides (Gongylus) ocellatus, Forskål.

17 specimens.

Some of these specimens have the pronounced coloration of the Berbera lizards, while others, so far as their colour is concerned, are in no way distinguishable from those from Egypt. Only in five out of the seventeen are there 28 rows of scales round the body, while twelve have 30, and one 32 rows of scales, thus overlapping the lepidosis of var. *tiligugu*.

At Maskat, the character of the coloration is similar to that just mentioned, but associated with it is a marked variation, in the relative development of the black and white spots, like that which occurs at Aden. In this variation the entire upper surface of the lizard is rich dark brown, and in place of the black spots brown ones are substituted, the white darts being reduced to fine points. The labials become nearly entirely brown, with a small white central spot. At Bushire, the typical form of colour is still present, but in intensity it resembles that of the Berbera, Aden, Hadramut, and Maskat lizards; but, as in the last, some specimens show a distinct tendency to assume the brown garb. At Jask, in Southern Persia, the South Arabian pronounced coloration is preserved, but, strange to say, one specimen from the same locality is pale greyish brown, while another is rich brown. These specimens have 30 rows of scales. In tracing the typical form to the east of Egypt, one is struck by its increase in size over those found in the Nile Valley, and in this respect the Arabian lizards also recall those found in Somaliland, but they have never the thick heavy bodies of var. *tiligugu*.

In the accompanying Table (p. 50) it will be seen that the variations that occur in the Hadramut lizards are very trifling as a whole.

e

Chalcides ocellatus, Forskål.

No. of specimen	Snout to vent	Tail	Eye to snout	Fore limb	Axilla to eye	Hind limb	Position of nostril	Position of postnasal	Supra-nasals	Upper and lower labials	Labials entering orbit	Scales round body
485	61	60	4·3	11·5	15	18	O. s.[1]	C. 1 & 2[2].	2	8·6[4]	5	30
486	75	77	5	15	18	21	″ ″	″ ″	2	9·6 : 8·6[5]	6	28
487	78	.....	5·7	16	19·2	22	″ ″	″ ″	2	8·6	5	28
488	81	84	5·8	16	20	24	″ ″	″ ″	2	8·6	5	30
489	86	95	6	16	21·6	24·7	″ ″	″ ″	2	8·6	5	30
484	89	.....	6	19	21	26	″ ″	″ ″	2	8·6	6	28
490	93	.....	6·4	19·5	21	26	′ ″	″ ″	2	9·6 : 9·6	R.6 : L.5	32
513	94	.....	6·3	17·5	22·5	26	″ ″	″ ″	2	9·7 : 8·7	5	28
491	98	.....	6·3	17·5	22·5	26·2	″ ″	″ ″	2	8·6	R.6 : L.5	30
492	102	.....	6·5	17·7	22	25·7	″ ″	″ ″	2	9·6 : 8·6	5	30
493	103	.....	7	19	24·2	27·4	″ ″	″ ″	2	8·6	5	30
483	103	102	7	20	24·3	27	″ ″	″ ″	2	8·6	5	28
494	105	.....	6·5	18·2	22·7	26·6	″ ″	″ ″	2	8·6	5	30
495	107	.....	6·7	18·6	24·3	27	″ ″	″ ″	2	8·6	5	30
496	110	.....	6	19·2	24·5	26	″ ″	R.O.0[3] : L.Cl	2	8·6	R.5 : L.3	30
497	112	.....	7	19·5	26·2	27·5	″ ″	C. 1 & 2	2	8·6 : 6·6	5	30
498	116	.....	7·2	21	26	26·4	″ ″		2	8·6	5	30

[1] Over suture of rostral and first labial.
[2] In contact with 1st and 2nd labial.
[3] Right side in contact with no labials.
[4] Eight upper and six lower labials.
[5] Nine upper labials on the right and six on the left side, and eight lower labials on the right and six on the left side.

RHIPTOGLOSSA.

CHAMÆLEONTIDÆ.

CHAMÆLEON CALCARIFER, Peters.

Chamæleon calcarifer, Anderson, Proc. Zool. Soc. 1895, p. 651.

4 ♀ and 9 ♂.

OPHIDIA.

COLUBRIDÆ.

ZAMENIS RHODORHACHIS, Jan.

2 ♂ and 2 ♀.

These four specimens agree with those from Aden in the low numbers of their ventrals as compared with Egyptian and Indian examples of the species.

Sex.	Total length.	Tail.	Ventrals.	Anal.	Caudals.	Scales.	Upper labials.	Labials entering orbit.	Relation of præ-ocular and frontal.	Postoculars.	Temporals.	Nasals.	Loreal.
♂...	1010	...	226	1/1	...	19	R. 9, L. 10	5 & 6	C.[1]	2	2+3	2	1
♂...	1167	297	227	1/1	129	19	R. 9, L. 9	5 & 6	B.C.[2]	2	2+3	2	1
♀...	780	215	228	1/1	125	19	R. 9, L. 9	5 & 6	B.C.	2	2+3	2	1
♀...	561	161	226	1/1	125	19	R. 9, L. 9	5 & 6	B.C.	2	2+3	2	1

Like the Aden specimens, they also belong to the variety *ladacensis*, with no vertebral stripe. They are greyish olive, one uniformly so, while the others have the characteristic dark cross-bands developed, chiefly anteriorly; and the angles of the ventrals are dusky, with a minute black spot; a dusky spot below the eye; underparts white.

ZAMENIS DIADEMA, Schleg.

1 ♂ and 1 ♀.

They have both 25 rows of scales round the body, which is a rare number, and the lowest that is found in this snake. I know of only two other instances of its occurrence, viz., in a specimen

---

[1] C. Contact.   [2] BC. Broadly in contact.

which my collector obtained at Duirat, Tunisia, and another in Mr. Blanford's collection from Karman, S.E. Persia. As a rule, the upper præocular is in contact with the frontal, but in both of these specimens it is excluded, except on the right side in the male. They have the usual markings.

TARBOPHIS GUENTHERI, Anderson.

Tarbophis guentheri, Anderson, Proc. Zool. Soc. 1895, p. 656, pl. xxxvi. fig. 3.

2 ♀.

The details of the external characters of these two specimens are given in the paper quoted above.

The specimen on which Forskål founded *Coluber dhara*[1] had a mutilated tail, with only 48 scutes; but the number of its ventrals, 235, and the description as a whole, suggest the possibility that *T. guentheri*, Anders., may be the same species. I think it, however, more probable that *C. obtusus*, Reuss, is *C. dhara*, Forskål, as the specimen from near Medina, which is not far off from Yemen, mentioned on p. 62 agrees with typical *C. obtusus*, Reuss, from Egypt. Unfortunately in Forskål's account there is no mention of the condition of the anal, and no information regarding the labials that entered the orbit.

CŒLOPELTIS MOILENSIS, Reuss.

Cœlopeltis moilensis, Reuss, Anderson, Proc. Zool. Soc. 1895, p. 656.

1 ♂ and 1 ♀.

Sex.	Total length.	Tail.	Ventrals.	Anal.	Caudals.	Scales.	Upper labials.	Labials entering orbit.	Præocular.	Relation of præocular and frontal.	Postoculars.	Temporals.	Nasals.	Loreal.
♂ ...	480	130	170	1/1	63	17	8	4 & 5	0	1 B. Ex.[2]	2	R. 2+2 / L. 2+2	1	1
♀ ...	487	129	176	1/1	73	17	8	4 & 5	0	1 B. Ex.	2	R. 2+2 / L. 2+3	1	1

[1] Descr. Animal. 1775, p. 14.   [2] Broadly excluded.

Psammophis schokari, Forskål.

Coluber schokari, Forskål, Descr. Anim. 1775, p. 14.
Coluber lacrymans, Reuss, Mus. Senck. i. 1834, p. 139.
Psammophis punctatus, Dum. & Bibr. Erpét. Génl. vii. 1854,
p. 896, Atlas, pl. 77, fig. of skull.
Psammophis sibilans, var. ?, Blyth, Journ. As. Soc. Beng. xxiv.
1855, p. 306.
Psammophis sibilans, var. *hierosolimitana*, Jan, Elenco, 1863,
p. 90 ; Icon. Génl., Livr. 34, Mars 1870, pl. iii. fig. 2.
Psammophis sibilans, var. *punctata*, Jan, Elenco, 1863, p. 90 ;
Boettger, Kobelt Reis. Alg. & Tun. 1885, p. 462.
Psammophis condanarus, var. *sindanus*, Stoliczka, Proc. As.
Soc. Beng. 1872, p. 83.
Psammophis leithii, Blanford, part., Eastern Persia, Zool. &
Geol. ii. 1876, p. 421 : part., Blgr. Trans. Linn. Soc. ser. 2, Zool.
v., 1889, p. 103 ; part., Fauna Brit. Ind., Rept. 1890, p. 365.
Psammophis moniliger, var. *hierosolymitana*, Boettger, Ber.
Senck. Nat. Ges. 1878-79, p. 65 ; id. loc. cit. 1879-80, p. 163.
Psammophis moniliger, var. *punctata*, Boettger, Ber. Senck.
Nat. Ges. 1879-80, p. 164.
Psammophis lacrymans, Boulenger, Proc. Zool. Soc. 1895,
p. 538 ; Anderson, *op. cit.* p. 655.

1 ♂, half-grown.

This specimen belongs to the variety in which the colour is
uniform, there being no longitudinal brown bands. It is pale
greyish olive above, becoming still paler on the tail. The upper
and lower labials, the throat, and the free margins of the ventrals,
on the anterior fourth of the snake, are dotted with blackish.
This type of coloration is found at Aden, in the Sinaitic Peninsula, and at Maskat. It occurs throughout Egypt, from Cairo
to Khartum, and at Durrur on the coast of the Red Sea. It is
also found in Persia, and as far east as Sind. Although I have
never met with the striped form in Egypt proper (Nile Valley),
it occurs at Suakin, which is close to Durrur, also at Maskat, in
Persia (Jask), and in Afghanistan. A similar coloration is also
essentially characteristic of the species in the extreme western
limit of its distribution, on the margin of the desert (Sahara),
in Tunisia, and Algeria.

It is generally found on the confines of the stony desert, with
which its colour is in unison, but, at Suakin, where it is more or

less striped, it occurs on the sandy and stony plain which is covered more or less with low thickets of bushes, and longish grass. It is extremely rapid in its movements.

In this male there are 170 ventrals and 141 caudals. The variation in the ventrals of this species may be as much as 32, the lowest number being 164, and the highest 195. In seven specimens from Arabia the variation is only 17; but if one of those from Maskat is excluded, the variation is only 11, the lowest number being 168 and the highest 179. This exceptional specimen has 185 ventrals, and in this high number it leads into the representatives of the species found in Persia, Baluchistan, Afghanistan, and Sind, which are almost invariably distinguished by a high number of ventrals ranging from 182 to 194. Among ten specimens from these countries, only one from Sind has its ventrals falling as low as 177. On the other hand, 16 specimens from Tokar, Suakin, and Durrur, have the ventrals varying from 163 to 174, but only in four does the number rise above 169, while in seven it does not exceed 166. The species, therefore, in the Suakin district, is characterized by a lower number of ventrals than in any other locality.

In the lower part of the Valley of the Nile, the ventrals vary from 168 to 177, but in Upper Egypt (Assuan) the number rises to 195, while, in the extreme west of its distribution, the high number 183 occurs at Biskra.

This species and *Psammophis leithii*, Gthr., have sometimes been mistaken the one for the other, and *Taphrometopon lineolatum*, Brandt, has occasionally not been distinguished from the latter. Their features are expressed by the following numbers:—

	Ventrals.	Anal.	Caudals.	Scales.	Upper labials.	Relation of praeocular and frontal.	Labials entering orbit.
P. sibilans.........	158–198	1/1	90–116	17	8	C.	4 & 5
P. schokari	163–194	1/1	93–131	17	9	B.C.	5 & 6
P. leithii	170–185	1	92–99	17	8	B.C.	4 & 5
T. lineolatum ...	175–194	1/1	72–90	17	9	B.C.	4, 5, & 6

VIPERIDÆ.

VIPERA ARIETANS, Merrem.

A young specimen measuring 210 millim. in length, of which the tail constitutes 14 millim. It has 32 rows of scales round the body, 136 ventrals, 1 anal, and 18 caudals.

This is the first record of the occurrence of this species in Asia.

ECHIS CARINATUS, Schn.

1 ♂.

ECHIS COLORATUS, Gthr.

3 ♀.

BATRACHIA.

ECAUDATA.

RANA CYANOPHLYCTIS, Schneider, Anders. Proc. Zool. Soc. 1895, p. 660, pl. xxxvii. fig. 2, tadpole.

A number of fine tadpoles.

PART III.

SOME REPTILES

FROM

OTHER PARTS OF ARABIA.

Reptiles from the Hejaz in the Cairo Museum.

THE following Reptiles from the Hejaz were collected for the Museum of the Medical School at Cairo, by one of its native employés : and I am indebted to Dr. Keatinge and to Dr. Walter Innes for having entrusted their identification to me.

PTYODACTYLUS HASSELQUISTII, Donndorf.

Le Gecko des Maisons, Cuv. Règn. Anim. ii. 1817, p. 49 ; Audouin, Descr. de l'Égypte, Nat. Hist. i. (1827), p. 165, Suppl. Rept. pl. i. figs. 2. 1 to 2. 6.

Lacerta gecko [1], Hasselquist & Linn. Iter Palæst. 1757, p. 306 : Linn. Mus. Lud. Ulr. Reginæ, 1764, p. 46 ; Syst. Nat. i. 1766 (part.), p. 365 : Forskål, Descr. Anim. 1775, p. viii et p. 13.

Stellio gecko, Schneider (part.), Amph. Phys. ii. 1792, p. 12.

Lacerta hasselquistii, Donndorf, Zool. Beytr. iii. 1798, p. 133.

Gekko ascalabotes, Merrem (part.), Tent. Syst. Amph. 1820, p. 40.

Gecko lobatus, Licht. Verz. Doubl. Berl. Mus. 1823, p. 103 ; Is. Geoffr. St. Hil. Descr. de l'Égypte, Nat. Hist. i. (1827), p. 132, pl. v. fig. 5.

Ptyodactylus lobatus, Gray, Ann. Phil. (2) x. 1825, p. 498 ; Fitz. Syst. Rept. 1843, p. 96 ; Boulenger, Cat. Liz. B. M. i. 1885, p. 110 ; Boulenger, Trans. Zool. Soc. xiii. 1891, p. 111, pl. xiii. fig. 2 ; Boettger, Kat. Rept. Mus. Senck. 1893, p. 27.

Ptyodactylus guttatus, Heyden, Rüppell's Atlas N. Afr., Rept. 1827, p. 13, pl. iv. fig. 1.

Ptyodactylus hasselquistii, Dum. & Bibr. Erpét Génl. iii. 1836, p. 378, pl. xxxiii. fig. 3 ; Rüppell, Mus. Senck. iii. 1845, p. 300 ; Gasco, Viaggio in Egitto, (pt. ii.) 1876, p. 110 ; Tristram, West. Palest., 1884, Rept. & Batr. p. 153 ; Hart, Fauna & Flora of Sinai,

---

[1] This name was first applied by Linnæus (Mus. Adolph. Frid. 1754, p. 46) to the Asiatic gecko, known as *G. verticillatus*, Laur.

1891, p. 210; Bœttger, Ber. Senck. Ges. 1879–80, p. 194; Boutan,
Rev. Biol. du Nord de la France, v. 1893,p. 336, fig. 1.

Ptyodactylus oudrii, Lataste, Le Natural.1880, p. 299; Boutan,
Rev. Biol. Nord France, v. 1893, p. 343, fig. 2.

Ptyodactylus lacazii, Boutan, Arch. Zool. Expér. (2) x. 1892,
p. 17.

Ptyodactylus bischoffsheimi, Boutan, Rev. Biol. Nord France,
v. 1893, p. 340, pl. iii. fig. 1.

Ptyodactylus montmahoui, Boutan, *l. c.* p. 369, pl. iii. fig. 2.

Ptyodactylus barroisi, Boutan, *l. c.* p. 375, pl. iii. fig. 3.

Ptyodactylus puiseuxi, Boutan, *l. c.* p. 379, pl. iii. fig. 4.

Ptyodactylus lobatus, subsp. *syriacus*, Peracca, Boll. Mus.
Torino, ix. 1894, no. 167, p. 1.

Ptyodactylus lobatus, var. *oudrii*, Werner, Verh. zool.-bot. Ges.
Wien, 1894, p. 76; Boettger, Zool. Centralblatt. June 1894, p. 376.

2 ♂. Hadir el Kabir near Medina.

1 juv. Dar Fadda between Medina and El Wish.

These two males present some resemblances in their general
form to the individuals of this species found on the plains of Suez,
but differ from them in their much more pointed snouts, in the
presence of enlarged tubercles on the thighs and on the tibiæ,
and in their tails being distinctly depressed. Two types of
nostril are met with in this species in the Nile Valley—one in
which it is so much raised above the snout as to merit the term
tubular being applied to it, and another in which the scales
defining the opening are only distinctly swollen. In the former
type, the nostril is surrounded by the first labial and three nasals,
while, in the simply swollen kind, it is enclosed by the rostral,
first labial, and three nasals. The individuals from the Hejaz
have nostrils of the latter type, but not so swollen as to entitle
them to be called semitubular. In the geckoes of the Plain of
Suez (2 specimens only), the nostrils are semitubular, and are
formed by the first labial and three nasals. Individuals in the
Nile Valley proper, with the nasal formula, rostral, labial, and
three nasals, have more or less depressed tails like the Hejaz
geckoes, while those in the former region, with the formula
L.3N., have rounded tails. There is, however, another important
character distinguishing them, and it is this, that the granules
of the body are more or less carinate while those on the snout
are distinctly keeled.

There is another feature, best marked in the adult male from Hadir el Kabir, less distinct in the other male, and absent in the young specimen, that calls for remark, as it is very rarely present in this species. It consists of the presence at the anterior border of the ear-opening of numerous spiny scales. A specimen, however, from the Mokattam hills, near Cairo, with the nasal formula R.L.3N., shows slight indications of the existence of similar scales, in the same position.

After a careful consideration of over sixty specimens of the geckoes of this genus from Palestine, the Dead Sea area, the Sinaitic Peninsula, the Hejaz, Maskat, the region between Shoa and Assab, the Nile Valley from Wádí Halfa to Suez, and from Algeria, I have arrived at the conclusion that the various modifications met with over the region indicated can be regarded in no other light than as illustrating the essentially polymorphic character of this species first described by Hasselquist from Lower Egypt.

AGAMA SINAITA, Heyden.

1 ♂. El Haggarieh between Medina and Mecca.

1 ♂. Near Medina.

The former has the following measurements :—Snout to vent 94; tail 172; length of head 24; width of head 22. Wrist in advance of the snout. Fourth toe reaches to anterior angle of eye. The scutes covering the bases of the claws, and the brown spines, on the under surfaces of the digits, are very well defined. Both have larger scales than specimens from the Nile Valley, and in this respect they resemble those from the Hadramut.

AGAMA RUDERATA, Olivier.

4 ♂ from near Medina.

These specimens are undoubtedly referable to this species. They have the characteristically enlarged scales on the body, limbs, and base of the tail. One has the head-scales nearly smooth, while, in the other, they are more or less convex. I have examined one of Olivier's specimens, from Arabia, preserved in the Paris Museum [1], with which these specimens practically agree, but the spines over the ears are more numerous and better developed than in Olivier's specimen, and the few spiny scales on

---

[1] It is preserved in the bottle numbered 2127, and the specimen itself bears a label no. 2610.

the back of the head and on the sides of the neck are more
strongly marked; but it must be borne in mind that Olivier's
lizard has not the freshness of these recently captured specimens, which are in excellent preservation. I have never met
with this species in Lower Egypt.

No.	Sex.	Snout to vent.	Tail.	Fore limb.	Hind limb reaches	Pores.	Tibia.	Head, length.	Head, breadth.	Locality.
159	♂	83	104	Fingers in advance of snout.	before ear.	28	19	22·2	21	Near Medina.
158	♂	81	...	,,	to ear.	33	19	21	21	,,
161	♂	79	100	,,	to ear.	20	18·6	20·5	20·2	Dar om Sheikh Ahmed, near Medina.
160	♂	76	104	,,	before ear.	21	18·3	20	20·4	,, ,,

AGAMA FLAVIMACULATA, Rüppell.

3 ♂ and 2 ♀. Near Medina.

Body moderately elongate, not depressed; head large, stoutly
and broadly cordate ; canthus rostralis short, not defined,
anterior to the nasal shield, which is circular with the nostril
directed upwards and backwards and perforated in the hinder
part of the shield internal to or on the line of the canthus rostralis; ear considerably smaller than the eye-opening, with a
fringe of pointed scales along its upper border; upper surface
of the head in adults covered with convex scales, especially large
along the mesial line ; a number of large, more or less obtusely
keeled scales behind the eye, more or less continuous with those
on the upper part of the post-temporal region, on which there
are a few short, strongly keeled, non-mucronate spiny scales.
Body covered above with moderately sized, unequal, imbricate,
more or less subacuminate, but not mucronate, feebly keeled
scales, those on the sides about half the size of the dorsal scales,
and more or less obtusely keeled; 77 to 95 scales round the
middle of the body ; ventral scales keeled or nearly smooth.
Limbs well developed, covered with regular keeled imbricate
scales of moderate size.   The wrist reaches to the anterior border
of the eye or to the nostril, and the tip of the fourth toe extends
to the ear or nearly so.   Tibia shorter than the skull in the

Agama flavimaculata, Rüppell.

No. of specimen.	Sex.	Snout to vent.	Tail.	Length of head.	Width of head.	Length of fore limb from head of humerus	Length of hind limb posteriorly.	Length of tibia.	Length of hind foot.	Pores.	Locality.
522.........	♂	77	107	18·3	21	42·5	55	18·8	23·5	0	The Hadramut, Arabia.
521.........	♀	60	80	15	15·5	34·2	43·4	14·2	18·2	0	,, ,,
B.M.	♂	60	77	14·8	16	32	43·1	14	16·3	0	Jiddah ,,
165.........	♂	117	150	29	29	57	74	24	32	0	Dar Asfan, between Medina and Mecca.
164.........	♂	103	129	25	26	52	70	22·5	30	0	,, ,,
166.........	♂	101	119	25·3	25	53	70	22·4	30	0	Near Medina.
169.........	♀	122	122	27	29	58	70	23	30	0	El Haggarieh, near Medina. Gravid.
167.........	♀	105	114	26	27	56	69	22	30	0	Near Medina, between Medina and Mecca.

adult. Tail longer or as long as the body and head, rounded, rather thick at the base especially in males, covered with regular, rather feebly keeled scales. A large gular pouch in both sexes. The usual gular and short præhumeral folds are present. Neither præanal pores nor callose scales are developed.

General colour pale yellowish on the head, olive-brown or even greyish brown on the body, generally many of the dorsal scales being yellow, but this character is occasionally absent. In adult males, the sides of the head, the gular pouch, the sides of the neck and body, are deep dusky brownish with a purple tinge, or rich bluish ou the sides of the mouth, gular pouch, sides of neck and shoulder, the tail being pale orange-yellow with occasionally deep orange spots along its sides. In some of these males, the general colour is pale olive-brown, the throat and chest being suffused with bluish, and the tail pale yellow. The adult breeding female is nearly olive, with many of the dorsal scales nearly white, arranged on the sides and on the limbs more or less in transverse series, especially on the limbs. The pouch is dark chocolate-brown, and the under surface of the head is marked with wavy lines of the same colour. Similar but more feeble markings occur on the chest.

The points in which this species differs from *A. jayakari,* Anders., are mentioned under the description of that species (p. 67).

The type was from Jiddah, and is preserved in the Frankfort Museum. I have examined it, and also the type of *A. leucostygma,* Reuss, to which Mr. Boulenger refers *A. flavimaculata,* Rüppell, as a synonym, a view which has also been adopted by Prof. Boettger.

The type of *A. flavimaculata,* Rüppell, is enumerated in Rüppell's Catalogue [1] as II. F.F. 6 *a–b,* 1834. The larger of the two specimens, a female, is the one figured by Rüppell, and both are stated in Prof. Boettger's [2] Catalogue to have come from Arabia, but Rüppell in his description [3] is more particular and gives Jiddah as the locality. There can be no doubt whatever regarding the identity of the Hejaz and Hadramut specimens with the type of the species, but the same cannot be said for their identity with *A. leucostygma,* Reuss. The types of the latter are two in number, and are marked Cat. II. F.F. 5 *k–l,* Gesch. 1827,

---

[1] Mus. Senck. iii. 1845, p. 302.
[2] Kat. Rept. Mus. Senck. 1893, p. 49.
[3] Neue Wirbelth. 1835, p. 12, pl. 6. fig. 1.

Dr. Rüppell, Ob. Ægypt. I cannot distinguish between them and the *A. pallida*, Reuss, the type of which I have also examined.

AGAMA STELLIO, Hasselq. & Linn.

One specimen from Mount Arafat in the Hejaz.

TARBOPHIS DHARA, Forskål.

Coluber dhara, Forskål, Descr. Animal. 1775, p. 14.
Coluber obtusus, Reuss, Mus. Senck. i. 1834, p. 137.
One ♂. Girenah Sidi Hamza near Medina.
This specimen agrees in all its details with those from Egypt, whence the species was described by Reuss. Forskål's *C. dhara* was from Yemen.
Total length 845 millim. Tail 123 millim. Ventrals 253. Anal 1/1. Caudals 60 ? Scales 23.

A Chameleon in the Cairo Museum.

I am indebted to Dr. Keatinge and to Dr. Walter Innes for having permitted me to bring to London, from the Cairo Museum, a very fine male specimen of a chameleon preserved in alcohol, and obtained, some years ago, in the Province of Yemen. And I have to record the obligation under which I lie to Professor Vaillant, of the Natural History Museum, Paris, for having fowarded to the British Museum one of the types of *C. calyptratus*, A. Duméril, in order that I might compare it with this male. The result of the comparison of the two has left no doubt in my mind regarding their specific identity.

CHAMÆLEON CALYPTRATUS, A. Duméril.

Cat. Méthod. Rept. 1851, p. 31 ; Arch. Mus. vi. p. 259, pl. xxi. fig. 1.

The types of this species are females, and they are said to have been obtained in the Nile Valley; but in the Catalogue of Reptiles in the Paris Museum, where the species was first indicated, no locality is given. The information that they came from the Nile Valley was supplied afterwards by A. Duméril. The specimen in the Cairo Museum from Yemen is the only example of the male in any Museum, so far as I am aware, and like the same sex of *C. calcarifer*, and other allied species, it has a tarsal spur.

There can be no doubt that this species is very closely allied to *C. calcarifer*, Peters, which is found at no great distance to the south of Yemen, viz. in the neighbourhood of Aden. The leading characters distinguishing the present species from *C. calcarifer*, Peters, are the great vertical elevation of the crest posteriorly, the little development of the occipital lobe, and the backward, but little upward, prolongation on to the crest of the supraorbital ridge. The crests and occipital lobes, however, of chameleons vary considerably, and *C. vulgaris*, Daud., is a very good illustration of this. In the other features the two are practically the same. The occurrence, therefore, of two so closely related forms in such near geographical relationship suggests the possibility, in view of the variations just mentioned, that, as the areas they inhabit are better explored, links connecting the one with the other may ultimately be discovered, but until this has happened the two must be kept distinct.

Reptiles from Aden collected by Captain C. G. Nurse.

Captain C. G. Nurse has lately presented four species of Reptiles from Aden to the British Museum, and I am indebted to Mr. Boulenger for permission to enumerate them here. They are the following:—*Hemidactylus yerburii*, Anders.; *Uromastix* (*Aporoscelis*) *benti*, Anders.; *Chamæleon calcarifer*, Peters; and *Glauconia nursii*, n. sp.

The *Hemidactylus* and *Chamæleon* call for no remark.

LACERTILIA.

Uromastix (Aporoscelis) benti, Anderson.

One ♂ and one juv.

These two lizards have a wonderful resemblance at first sight to *Uromastix ornatus*, Heyden, but they are at once distinguished from it by the absence of præanal and femoral pores. In those species which have these structures they are always present even in very young individuals, but although in these two specimens there are no traces of them, both have a line of enlarged scales along the thighs in the position occupied by these structures, and more or less callose, but structurally quite distinct from the deep pit of a true femoral pore. In the typical form of this species there is not the same clear definition of these femoral scales as that present in the Aden specimens, but as the latter also present other slight modifications in their lepidosis, as compared with the types, the differences in question I ascribe wholly to local variation. These Aden specimens have somewhat smaller scales, which is shown by the number round the body being as high as 209 in one and 190 in the other, whereas in the typical form the variation is from 146 to 176.

They were obtained from the hills 50 miles from Aden.

	♂.	Juv.
Total length	271	146
Tail	124	65
Length of head	27	16
Width of head	27	16

OPHIDIA.

GLAUCONIA NURSII, n. sp.

Two specimens.

Head rather broad; neck distinctly narrower than the head; snout rounded; rostral broader than the nasal, prolonged backwards nearly on a level with the eyes; nasal completely divided; nostril removed from the rostral; first labial very small, slightly higher than broad, its upper and lower margins nearly parallel, and its breadth slightly less than that of the lower part of the nasal; ocular in the labial margin; supraocular and frontal shields nearly equal. Diameter of the body about fifty times in the total length; tail ten times in the length. Two hundred and eighty-one scales from the labial margin to the vent, and thirty-six on the tail; fourteen scales round the body. Pale brownish above, nearly white below.

This species is most closely allied to *G. blanfordii*, Blgr., but is at once distinguished from it by its much narrower rostral, and by its shorter first labial which is as broad as high, and about equal to the breadth of the lower portion of the nasal; its body also is somewhat stouter, and the head is broader and better defined.

Total length	250	230
Tail	25	20
Diameter of body at middle	5	4·5

A new Agamoid Lizard from Maskat.

In going over the species of the genus *Agama* added to the British Museum collection since the appearance of the first volume of Mr. Boulenger's 'Catalogue of Lizards,' I found the following species from Maskat to be one hitherto unrecognized, and I have Mr. Boulenger's permission to describe it.

AGAMA JAYAKARI, n. sp.

Agama isolepis (*not* Boulenger), Boulenger, Ann. & Mag. N. H. (5) xx. 1887, p. 407 [1].

Body elongate, not depressed ; head large, more or less triangularly cordate ; snout moderately pointed ; canthus rostralis short, not defined anterior to the nasal shield, which is circular with the nostril directed upwards and backwards, and perforated in the hinder part of the shield internal to or on the canthus rostralis ; ear considerably smaller than the eye-opening, with a fringe of pointed scales along its upper border. Upper surface of the head, in adults, with convex scales, especially large along the mesial line ; scales on the temporal region large, generally well keeled, but sometimes nearly smooth ; scales above, below, and behind the ear and on the occipital region with sharp, prominent, mucronate points. Body covered above with large, equal, strongly keeled, mucronate scales gradually diminishing in size on the sides, where they are about half the size of the dorsal scales and without keels, but furnished with sharp points; about 90 scales around the middle of the body ; ventrals strongly or moderately keeled. Limbs well-developed, covered with regular, moderately large, strongly keeled, slightly mucronate scales ; the wrist generally reaches the nostril, but occasionally to the snout, and the tip of the fourth toe extends to the ear, or it may fall somewhat short of it. Tibia longer than the skull. Tail considerably longer than the body and head, rounded, rather thick at the base, especially in the males, and covered with strongly keeled imbricate scales nearly as large as the dorsal scales. A large gular pouch in both sexes, with the usual gular and short præhumeral fold. Neither præanal pores nor callose scales are present.

General colour of the upper parts olive suffused with brownish,

---

[1] When Mr. Boulenger made this identification, he had only one specimen before him, and that a female.

f

Agama jayakari, n. sp.

No. of specimen.	Sex.	Snout to vent.	Tail.	Length of head.	Width of head.	Length of fore limb from head of humerus.	Length of hind limb posteriorly.	Length of tibia.	Length of hind foot.	Pores.	Locality.
94. 3. 21. 4 ...	♂	138	178	28	28	64	91	30	38	0	Maskat.
5 ...	♂	133	176	25·8	31	73	98	31	41	0	,,
6 ...	♂	125	185	26·2	29	66	86	29	36·2	0	,,
7 ...	♂	125	187	25·2	27	68	90	30·8	39	0	,,
8 ...	♂	118	179	......	29	69	85	28	34	0	,,
87. 11. 11. 12 ...	♀	115	151	22·5	25	60·5	72·7	24·8	33	0	,,
94. 3. 21. 4 ...	♀	108	141	21·5	24·6	55	76	24·8	31·3	0	,,

or with bluish green, and more or less spotted with yellowish, and the limbs and tail with brownish. The head generally bluish green and yellow; on the back of the neck there are usually indications of two short longitudinal brownish bands.

This species is closely allied to *A. flavimaculata*, Rüppell, from Jiddah, but is distinguished from it by its large, regular, strongly keeled and mucronate scales, by its less cordate head, which is shorter than the tibia, and by the more strongly keeled character of its caudal scales. It also attains to a greater size, and the scales on the back of the head are somewhat more spinose than in *A. flavimaculata*, Rüppell. The entire absence of præanal pores and callose scales in these two species necessitates an alteration in the hitherto accepted definition of the genus *Agama*.

A. isolepis, Blgr., which has præanal pores, is more allied to *A. agilis*, Oliv., than to this species.

Judging from the number of specimens sent to the British Museum by Dr. Jayakar, it would appear to be well represented at Maskat. It attains to a greater size than *A. sanguinolenta*, Pallas.

PART IV.

SKETCH OF THE LITERATURE

BEARING ON THE

REPTILIAN AND BATRACHIAN FAUNA OF ARABIA.

As a supplement to these lists of Reptiles, I have now to add a complete list of all the species of Reptilia and Batrachia known to occur in Arabia, and as an introduction to it I give a short sketch of the literature of the subject.

The Danish Expedition equipped by Frederick V. of Denmark for the scientific exploration of Egypt, Arabia, and Syria, sailed from Europe in January 1761. When it had accomplished its work in Egypt, the members of the Expedition went to Suez, and after Niebuhr had visited Mount Sinai they sailed on the 21st July, 1762, for Jiddah, whence they proceeded overland to Mocha. Here Van Haven, the philologist, died ; and two months afterwards Peter Forskål, the naturalist, succumbed to the plague, at Jerim, on the 11th July, 1763. Shortly after this, Niebuhr left Arabia, but he did not return to Copenhagen until November, 1767, when he at once began to prepare for publication the results achieved by the Expedition, and to arrange, for the same purpose, the materials that his friend Forskål had brought together. This posthumous work of Forskål's, entitled 'Descriptiones Animalium, &c.,' appeared in 1775. It is our first introduction to the fauna of Arabia.

In it he mentions the occurrence in Loheia, north of Hodeida, of a land-tortoise, *Testudo terrestris*, and the local name of which is *Buzi* or *Sukar*. It is useless attempting to identify this animal, but it may possibly be *T. leithii*, Günther, which is found in the desert country between Ismailia and El Arisch ; while, on the other hand, it may prove to be *Testudo elegans*, Schoepff, which has been recorded from Muscat.

Four species of Lizards are described, viz. : —*Lacerta nilotica*, Hasselquist & Linnæus, = *Varanus niloticus*, Hasselq. & Linn. ; *Lacerta ægyptia*, Hasselquist & Linnæus, = *Uromastix ægyptius*, H. & L. ; *Lacerta ocellata*, Forskål, = *Chalcides (Gongylus) ocellatus*, Forsk. ; and *Lacerta gecko*, H. & L., = *Ptyodactylus hassel-*

quistii, Donndorf. There is no explicit statement whence these species were obtained. The first-mentioned is certainly not found in Arabia; while, on the other hand, all of them are present in Egypt, and since Hasselquist's day have been recorded from Arabia with the exception of the Nile Monitor.

Forskål describes eight species of Snakes:—1. *Coluber lebetinus*; 2. *Coluber guttatus*, Forskål; 3. *Coluber haje*, Forskål; 4. *Coluber dhara*, Forskål; 5. *Coluber schokari*, Forskål; 6. *Coluber bœtan*, Forskål; 7. *Coluber hölleik*, Forskål; and 8. *Coluber* ——, Arab. *Hannasch asuœd*, or black snake. The first was received from Cyprus; the second is assigned to Cairo, and is probably the species afterwards described by Is. Geoffroy St. Hilaire as *C. florulentus = Zamenis florulentus*; the third is *Naia haje*, but the place of its occurrence is not stated; the fourth is evidently a member of the genus *Tarbophis*, and the snake described as *C. obtusus*, Reuss, is, I believe, identical with it, seeing that the Hejaz specimen in no way differs from the Egyptian snake described by Reuss; the fifth is unquestionably the *Psammophis* redescribed in after years under a variety of names, e. g., *P. lacrymans*, Reuss, *P. punctatus*, D. & B., and *P. sibilans*, var. *hierosolimitana*, Jan, &c. The descriptions of the three remaining species are too vague to enable them to be determined. Only two of the eight species, viz. *Tarbophis dhara* and *Psammophis schokari*, are ascribed to Arabia, and to the Province Yemen.

Besides these reptiles, Forskål mentions a number of others, under their native names, chiefly, from Syria, Egypt, and Arabia, but it would be vain to attempt to identify them.

Olivier, in the beginning of the century, referred a lizard from North Arabia to *A. ruderata*. This specimen is preserved in the Paris Museum [1].

Rüppell, about the middle of the second decade of this century, began his exploration of North-Eastern Africa, and in the course of his travels visited Arabia Petræa, the Sinaitic Peninsula, and the ports of Moilah and Jiddah. The results of his first journey were made known by Dr. C. H. G. Heyden, in 1827. In this work the following species of Reptiles and Batrachia are described from Arabia, viz.:—*Uromastix ornatus*, Heyden; *Agama sinaita*, Heyden; *Agama stellio*, H. & L.; *Ptyodactylus guttatus*, Heyden, =*P. hasselquistii*, Donndorf; *Stenodactylus scaber*, Heyden,=

---

[1] Cat. Rept. Paris Mus., Duméril, 1851, p. 103.

Gymnodactylus scaber, Heyden ; *Hemidactylus granosus*, Heyden, = *Hemidactylus turcicus*, Linn. ; *Bufo arabicus*, Heyden, = *Bufo viridis*, Laur.

Dr. A. Reuss, in 1834, characterized a number of reptiles from Arabia collected by Rüppell, principally on his second expedition. Oue, however, was obtained on his first journey, and is stated to have been found at Tor, Arabia Petræa, a locality I have not been able to identify, unless it be Tor, on the sea-coast of the Sinaitic Peninsula. This species, viz. *Lacerta longicaudata* = *Latastia longicaudata*[1], was afterwards found by Rüppell in Abyssinia, where it appears to be common, as it is likewise at Suakin, in the neighbourhood of which I obtained many specimens. Reuss also enumerates the following species as Arabian :— *Agama loricata*[2] = *A. pallida*, Reuss[2]; *Coluber lacrymans* = *Psammophis schokari*, Forskål ; and *Coluber moilensis* = *Cœlopeltis moilensis*, Reuss.

Rüppell himself, in 1835, described *Trapelus flavimaculatus* = *Agama flavimaculata*, from Jiddah, to which locality Duméril and Bibron erroneously ascribed *Agama cyanogaster*, Rüppell, which was first discovered at Massowah.

In the fourth volume (1837) of Duméril and Bibron's work, *Agama flavimaculata*, Rüppell[3], is wrongly regarded as a synonym of *A. agilis*, Olivier[4], and consequently Jiddah is given as a locality for the latter species.

In 1837, Wiegmann described *Scincus meccensis* from Mecca ; and, in 1839, Dr. J. E. Gray referred a lizard collected by Rüppell in Arabia Petræa to the genus *Riopa*, but now generally recognized to be *Ablepharus pannonicus*, Fitz.

In Rüppell's Catalogue of the Reptiles in the Frankfort Museum, a specimen of *Stenodactylus guttatus*, Cuv., = *S. elegans*, Fitz., is mentioned from Arabia.

Duméril and Bibron, in their sixth volume (1844), record *Typhlops vermicularis*, Merr., from the foot of Sinai, on the strength of specimens from that locality in the Leyden Museum, and also note the presence of *Eryx jaculus*, Hasselq. & Linn., in Arabia, but do not state the source of their information regarding the latter.

[1] Prof. O. Boettger (Kat. Rept. Frankfort Mus. 1893, p. 89) records this species only from Arabia, but besides the type from Tor there are Rüppell's three Abyssinian specimens.

[2] Type in the Frankfort Museum examined.

[3] Type examined.

[4] Types examined.

In A. Duméril's Catalogue of the Reptiles in the Paris Museum (1851), *Agama cyanogaster*, Rüppell, *Euprepes septemtæniatus*, Reuss,=*Mabuia septemtæniata*, Reuss, and *Gongylus ocellatus*, Forskål,=*Chalcides* (*Gongylus*) *ocellatus*, Forskål, are represented by Arabian specimens.

In the seventh volume of the ' Erpétologie Générale ' (1854), *Psammophis punctulatus*, D. & B., is described from Arabia, whence it had been obtained by M. Arnaud, who travelled as a collector for the Paris Museum.

In 1863, Dr. J. E. Gray described *Spatalura carteri*=*Pristurus carteri*, a rock-gecko found by Mr. H. J. Carter on the island of Masira, and mentioned the existence at Makallah of a *Uromastix* which he was unable to identify. In the following year he described a chameleon from Arabia, *O. auratus*=*C. vulgaris*, Daud.

Professor Peters, in the following year, recorded *Chalcides* (*Sphænops*) *sepoides*, Aud., from Tor, Arabia, under the name of *Sphænops capistratus*, Wagler.

Strauch, in his Monograph of the Viperidæ (1869), mentioned that a specimen of *Cerastes cornutus*, H. & L., from Arabah, Arabia Petræa, was preserved in the Munich Museum ; and in the same year Westphal-Castlenau gave Arabia as one of the localities in which *E. pardalis*, D. & B.,=*Eremias guttulata*, Licht., was found.

In 1871, Peters described *Pristurus longipes*=*Pristurus crucifer*, Val., from Aden ; and in that year I characterized a skink, *Scincus mitranus*, from Arabia, allied to *S. officinalis*, Schn.

In 1874, Mr. Blanford gave an account of a new gecko from Maskat, *Pristurus rupestris*, and recorded *Pristurus flavipunctatus*, Rüppell, from the same locality.

Captain, afterwards Sir Richard, Burton presented, in 1878, to the British Museum a small collection of reptiles that he had brought together in Midian. The species were described and identified by Dr. Günther, who recognized two as new to science, viz. *Zamenis elegantissimus* and *Echis coloratus* [1] ; while the following were new to the Arabian fauna:—*Zamenis cliffordi*= *Z. diadema*, Schlegel ; *Zamenis ventrimaculatus*=*Z. rhodorhachis*, Jan ; *Echis carinatus*, Schn. ; *Ceramodactylus doriæ*, Blanford, =*Stenodactylus* (*C.*) *doriæ*, Blanf. ; *Uromastix spinipes*, Merrem,= *U. ægyptius*, Hasselquist & Linn. ; *Acanthodactylus boskianus*, Daud. ; *Acanthodactylus cantoris*, Günther,=*A. boskianus*, Daud. ; and *Bufo pantherinus*, Boie,=*B. regularis*, Reuss.

[1] Recorded from the island of Socotra by Dr. Günther, P. Z. S. 1881, p. 463.

Dr. J. v. Bedriaga, in his Catalogue of Amphibia and Reptilia of Western Asia (1879), stated that *Zamenis karelinii*, Brandt, had been found at Cape Massendam, and mentioned the occurrence of *Cœlopeltis lacertina*=*C. monspessulana*, Hermann, in Arabia; but the authorities on which these statements were made are not given [1].

Prof. O. Boettger, in 1879–80, recorded *Platydactylus mauritanicus*, Linn., = *Tarentola mauritanica*, Linn., from Arabia on the strength of a verbal communication from Dr. E. Buck.

In 1882, Professor Vaillant stated that the Paris Museum had received a specimen of *Uromastix princeps*, O'Shaughn., from Aden, but at the same time suggested that it had been taken there accidentally. This is probable; and as it is a striking form, owing to its short, broad, spiny tail, it may have been purchased from some Somali or other native who had taken it from Berbera to Aden, in the expectation of selling it. On the other hand, in view of the number of species common to the two sides of the Straits, it is just possible that this species may ultimately be found in Arabia, but as its presence there is doubtful, I have not included it in the accompanying list.

Dr. Lortet, in his Catalogue of Syrian Reptiles (1883), mentioned the occurrence of a land-tortoise between Ismailia and El Arisch under the name of *Testudo kleinmanni*, Lortet, = *T. leithii*, Gthr.

In 1885, Mr. Boulenger, in the first volume of his 'Catalogue of Lizards,' described a new gecko from the Sinaitic peninsula as *Hemidactylus sinaitus*, and recorded *Tarentola annularis*, Is. Geoffr., from the same region. He also identified the *Uromastix* from Makallah, which Dr. Gray had been unable to name, as *U. hardwickii*, Gray.

Mr. J. A. Murray, in the following year, made known a new skink, *S. muscatensis*, from Maskat, allied to *S. mitranus*, Anders.

In 1887, Mr. Boulenger published a list of Reptilia and Batrachia from Maskat, collected by Dr. A. S. G. Jayakar. The following species were added to the fauna of Arabia:—*Testudo stellata*, Schn., = *T. elegans*, Schoepff, possibly introduced; *Alsophylax tuberculatus*, Blanf., = *Bunopus tuberculatus*, Blanf.; *Hemi-*

---

[1] It is just possible that Dr. Bedriaga may have had in view the snake mentioned by Rüppell as a variety of *C. monspessulana*, Hermann, which appears to be *C. moilensis*, Reuss. As far as I have been able to make out, *C. monspessulana* has only recently been recorded by Mr. Hart as inhabiting the Arabian Province.

dactylus coctæi, D. & B., = *H. flaviviridis*, Rüppell ; *Agama isolepis*, Blgr., = *A. jayakari*, Anders. ; *Varanus griseus*, Daud. ; *Lacerta jayakari*, Blgr. ; *Lytorhynchus diadema*, D. & B. ; *Dipsas obtusa*, Reuss, = *Tarbophis guentheri*, Anders. ; and *Bufo andersoni*, Blgr. In the same year, Mr. Boulenger, in the third volume of his ' Catalogue of Lizards,' mentioned the following species as represented in Arabia :—*Acanthodactylus scutellatus*, D. & B., *Eremias mucronata*, Blanf., *Eremias rubropunctata*, Licht., *Mabuia quinquetæniata*, Licht., and *Chamæleon calcarifer*, Peters ; and, in 1888, he described a new *Eryx* from Maskat, *E. jayakari*.

Mr. H. C. Hart, in his ' Fauna and Flora of Sinai, Petra, and Wádí Arabah ' (1891), recorded *Rhynchocalamus melanocephalus*, Gthr., = *Oligodon melanocephalus*, Gthr., from Petra half-way between Akabah and the Dead Sea, and *Rana esculenta*, Linn., from Ghôr, at the southern extremity of the Dead Sea.

Prof. O. Boettger, in 1892, added *Scincus hemprichii*, Wiegm., to the Arabian fauna.

In the following year, Mr. P. Matschie described a new species of *Latastia* as *Philochortus neumanni* = *L. neumanni*, from Aden, and mentioned the presence in the same locality of *Mabuia pulchra*, Matschie, = *Mabuia brevicollis*, Wiegm., *Rana ehrenbergi*, Peters, = *R. cyanophlyctis*, Schn., and *Bufo arabicus*, Matschie, = ? *Bufo pentoni*, Anders.

In 1894, I described two new lizards from the district between Makallah and the Hadramut, viz. :—*Uromastix (Aporoscelis) benti* and *Phrynocephalus arabicus*; and, in the following year, added two new lizards from Aden, *Hemidactylus yerburii* and *Mabuia tessellata*, and pointed out that the Muscat snake identified by Mr. Boulenger as *Dipsas obtusa*, Reuss, is a distinct species which I named *Tarbophis guentheri*. I also recorded the presence of *Acanthodactylus cantoris*, Gthr., and *Bufo pentoni*, Anders., at Aden.

In the foregoing list of Reptiles and Batrachia, collected on Mr. Bent's Expedition, the following are additions to the fauna of Arabia, viz. :—*Stenodactylus (C.) pulcher*, n. sp., *Bunopus blanfordii*, Strauch, *Pristurus collaris*, Steindachn., *Agama adramitana*, n. sp., *Eremias brevirostris*, Blanf., *Scincus conirostris*, Blanf., and the puff-adder, *Vipera arietans*, Merrem.

To this list has to be added the new Agama from Maskat, *Agama jayakari*, the chameleon from Yemen, *C. calyptratus*, A. Duméril, and the *Glauconia* from Aden, *G. nursii*, Anders.

The foregoing species enumerated from Arabia number seventy-nine in all, excluding the two species of *Rana* and the four species of *Bufo*.

They represent one Chelonian genus, twenty Lacertilian genera, and twelve genera of Ophidia.

The species are distributed thus :—Testudinidæ, 2 species ; Geckonidæ, 18 species ; Agamidæ, 13 species ; Varanidæ, 1 species ; Lacertidæ, 10 species ; Scincidæ, 12 species ; Chamæleontidæ, 3 species ; Typhlopidæ, 1 species ; Glauconiidæ, 1 species ; Boidæ, 2 species ; Colubridæ, 12 species ; Viperidæ, 4 species. Making in all 2 species of land-tortoises ; 57 species of lizards ; and 20 species of snakes.

The following are the genera and the number of species representing each genus :—

CHELONIA.

Testudo .... 2 species.

SQUAMATA.
LACERTILIA.

GECKONIDÆ.		LACERTIDÆ.	
Stenodactylus	3 species.	*Lacerta*	1 species.
Bunopus	2 species.	*Latastia*	2 species.
Gymnodactylus	1 species.	*Acanthodactylus*	3 species.
Pristurus	5 species.	*Eremias*	4 species.
Ptyodactylus	1 species.		—
Hemidactylus	4 species.		10
Tarentola	2 species.		
	—	SCINCIDÆ.	
	18	*Mabuia*........................	4 species.
AGAMIDÆ.		*Ablepharus*	1 species.
Agama	8 species.	*Scincus*	5 species.
Phrynocephalus	1 species.	*Chalcides*	2 species.
Uromastix	4 species.		—
	—		12
	13		
VARANIDÆ.		CHAMÆLEONTIDÆ.	
Varanus	1 species.	*Chamæleon* ....:....	3 species.

OPHIDIA.

TYPHLOPIDÆ.		COLUBRIDÆ.	
Typhlops	1 species.	*Zamenis*	4 species.
GLAUCONIIDÆ.		*Lytorhynchus*	1 species.
Glauconia	1 species.	*Oligodon*	1 species.
BOIDÆ.			—
Eryx...........................	2 species.		6

OPHIDIA (*cont.*).

COLUBRIDÆ (*cont.*) ...	6		VIPERIDÆ.	
Tarbophis	2 species.		*Vipera*	1 species.
Cælopeltis	2 species.		*Cerastes*	1 species.
Psammophis................	2 species.		*Echis*	2 species.
	12			4

From the succeeding Lists of Arabian Reptiles and the analyses accompanying them, it will be seen that 21 species are found in Arabia and nowhere else. Leaving these out of consideration, and also *Hemidactylus flaviviridis*, Rüppell, *Zamenis diadema*, Schl., and *Echis carinatus*, Schn., all of which have a wide distribution, embracing the greater part of India and North Africa, 55 species remain, and of these only 13 are not found in Africa; so that the fauna has a most marked African character. Of the remaining 42 species, 23 are found in the Nile Valley, viz.:—*Testudo leithii*, Gthr., *Stenodactylus elegans*, Fitz., *Bunopus blanfordii*, Strauch, *Gymnodactylus scaber*, Heyden, *Ptyodactylus hasselquistii*, Donndorf, *Tarentola annularis*, Is. Geoffr., *Agama sinaita*, Heyden, *Agama pallida*, Reuss, *Uromastix ægyptius*, Hasselq. & Linn., *Varanus griseus*, Daud., *Acanthodactylus boskianus*, Daud., *Acanthodactylus scutellatus*, Aud., *Eremias guttulata*, Licht., *Eremias rubropunctata*, Licht., *Mabuia quinquetæniata*, Licht., *Chalcides (S.) sepoides*, Aud., *Chamæleon calyptratus*, A. Dum., *Zamenis rhodorhachis*, Jan, *Lytorhynchus diadema*, D. & B., *Tarbophis dhara*, Forsk., *Cælopeltis moilensis*, Reuss, *Psammophis schokari*, Forsk., and *Cerastes cornutus*, Hasselq. & Linn. The affinity, however, of the North-Western portion of the Arabian fauna with that of Lower Egypt is further manifested by the presence in it of the species which enter the Nile Valley from the north, viz.:—*Agama stellio*, Hasselq. & Linn., *Eryx jaculus*, Hasselq. & Linn., and *Oligodon melanocephalus*, Jan, and by the occurrence of the Mediterranean species *Hemidactylus turcicus*, Linn., *Tarentola mauritanica*, Linn., *Chalcides (G.) ocellatus*, Forsk., *Chamæleon vulgaris*, Daud., and *Cælopeltis monspessulana*, Hermann; so that if these be added to the 23 already enumerated, 31 species, so far, are common to Egypt and Arabia. But the affinity of the two faunas is still further emphasized by a consideration of the species that are distributed between the Nile and the Red Sea. The following species occur in the Southern end of that area, viz.:—*Pristurus flavipunctatus*, Rüppell, *Pristurus crucifer*, Val., *Hemidactylus sinaitus*, Blgr., *Mabuia brevicollis*,

Wiegm., *Mabuia septemtæniata*, Reuss, and *Scincus hemprichii*, Wiegm.; and all of them are found on the eastern side of the Straits of Bab el Mandeb. *Agama cyanogaster*, Rüppell, is also an Abyssinian and Somaliland lizard that occurs on the Arabian coast, at Jiddah; and to it have to be added two other Abyssinian and Eastern Sudan species, viz., *Latastia longicaudata*, Reuss, and *Eremias mucronata*, Blanf., which, strange to say, have not hitherto been found in South Arabia, but only in the Sinaitic Peninsula. *Psammophis punctulatus*, D. & B., another species inhabiting Abyssinia, has also been recorded from Arabia, but from what part has not been stated. *Vipera arietans*, Merr., has recently been received from Somaliland, and it is now for the first time recorded from Asia. If to the foregoing 42 species, *Hemidactylus flaviviridis*, Rüppell, *Zamenis diadema*, Schlegel, and *Echis carinatus*, Schn., be added, the list of 45 species occurring on both sides of the Red Sea and of the Suez Canal is completed. *Ablepharus pannonicus*, Fitz., *Typhlops vermicularis*, Merr., and *Agama ruderata*, Olivier, found in the north-western extremity of Arabia, do not enter Africa; whilst *Echis coloratus*, Gtbr., is found in Socotra and Palestine. Two species—*Stenodactylus (C.) doriæ*, Blanf., and *Scincus conirostris*, Blanf.,—found in South Persia, and not to the west of the Persian Gulf, occur also in Eastern Arabia, along with *Zamenis karelinii*, Brandt, which ranges southwards to Persia from Turkestan. In the same part of Arabia *Testudo elegans*, Schoepff, and *Uromastix hardwickii*, Gray, both of which are essentially Indian forms, are likewise present; and associated with them are other four species, viz.:— *Bunopus tuberculatus*, Blanf., *Pristurus rupestris*, Blanf., *Acanthodactylus cantoris*, Gthr., and *Eremias brevirostris*, Blanf., all present in the north-western portion of India.

Two frogs and four toads have been observed in Arabia. Two are European species, viz., *Rana esculenta*, Linn., and *Bufo viridis*, Laur.; the former extends into North-Western, and the latter into North Africa generally. Two of the toads, viz., *Bufo regularis*, Reuss, and *Bufo pentoni*, Anders., are closely allied, and are confined to Africa and Arabia, the former being only found in North-Western and the latter in Southern Arabia. The two remaining species, *Rana cyanophlyctis*, Schn., and *Bufo andersoni*, Blgr., are confined to Asia, the former having a very wide range from Malaya to S.E. Arabia, while the latter is a North-West Indian species extending into S.E. Arabia.

PART V.

LIST OF THE

REPTILIA AND BATRACHIA OF ARABIA,

1775 *to* 1896.

REPTILIA.

CHELONIA.

Testudo elegans, Schœpff [1], 1792–1801.
 Maskat (*Jayakar*), Boulenger, 1878.
Testudo leithii, Gthr.[2], 1869.
 Between Ismailia and El Arisch, Lortet, 1883.

SQUAMATA.
LACERTILIA.

Stenodactylus elegans, Fitz., 1826.
 Arabia, Rüppell, 1843–45 ; Sinai, Sinaitic Peninsula (*Hart*),
 Boulenger, 1885 ; Mount Sinai, Hart, 1891.
Stenodactylus (Ceramodactylus) pulcher, Anders., 1896.
 Hadramut, Anderson, 1896.
Stenodactylus (Ceramodactylus) doriæ, Blanf., 1874.
 Arabia (*Sir R. Burton*), Günther, 1878 ; Sinaitic Peninsula
 (*Hart*), Boulenger, 1885 ; Aden (*Yerbury*), Anderson,
 1895 ; Hadramut, Anderson, 1896.
Bunopus tuberculatus, Blanf., 1874.
 Maskat (*Jayakar*), Boulenger, 1887.
Bunopus blanfordii, Strauch, 1887.
 Hadramut, Anderson, 1896.
Gymnodactylus scaber, Heyden, 1827.
 Stony plains, Tor, Arabia (*Rüppell*), Heyden, 1827 ; Aden
 (*J. Schmidt*), Boettger, 1892 ; Maskat (*Bornmüller*),
 Werner, 1894.
Pristurus flaripunctatus, Rüppell, 1835.
 Maskat, Blanford, 1874 ; Aden (*Yerbury*), Anderson, 1895.

---

[1] Possibly imported.
[2] Described from a Sind specimen, probably imported.

Pristurus rupestris, Blauf., 1874.

Maskat, Blanford, 1874; Hadramut, Anderson, 1896.

Pristurus crucifer, Valenciennes, 1861.

Kursi, near Aden (*Marquis Doria*), Peters, 1871.

Pristurus collaris, Steindachner, 1867.

Hadramut, Anderson, 1896.

Pristurus carteri, Gray, 1863.

Island of Masira (*Carter*), Gray, 1863.

Ptyodactylus hasselquistii, Donndorf[1].

Tor, Arabia Petræa, Sinai (*Rüppell*), Heyden, 1827; Sinaitic Peninsula (*Hart*), Mt. Sinai, Boulenger, 1885; Maskat (*Jayakar*), Boulenger, 1887; Sinai and Arabah, Hart, 1891; Caverns of Hammam Far 'un, 60 miles S.E. of Suez, Boutan, 1892; Medina District, Hejaz, Anderson, 1896.

Hemidactylus sinaitus, Blgr., 1885.

Mt. Sinai, Boulenger, 1885; Aden (*Yerbury*), Anderson, 1895; Hadramut, Anderson, 1896.

Hemidactylus turcicus, Linn., 1766.

Arabia (*Rüppell*), Heyden, 1827; Arabia Petræa, Rüppell, 1843–45; Sinai, Werner, 1893; Hadramut, Anderson, 1896; Tor, Sinaitic Peninsula, Anderson, 1896.

Hemidactylus yerburii, Anderson, 1895.

Aden (*Yerbury*) (*Nurse*), Anderson, 1896.

Hemidactylus flaviviridis, Rüppell, 1835.

Maskat (*Jayakar*), Boulenger, 1887; Aden (*J. Schmidt*), Boettger, 1892; Aden, (*Neumann*) Matschie, 1893, (*Yerbury*) Anderson, 1895; Hadramut, Anderson, 1896.

Tarentola annularis, Is. Geoffr., 1827.

Mount Sinai, Boulenger, 1885.

Tarentola mauritanica, Linn.[2], 1766.

Arabia (*Buck*), Boettger, 1879–80.

Agama sinaita, Heyden, 1827.

Sinai (*Rüppell*), Heyden, 1827; Maskat, A. Duméril, 1851; Sinaitic Peninsula (*Hart*); Mt. Sinai, Boulenger, 1885; Maskat (*Jayakar*), Boulenger, 1887; Sinai and Akabah, Hart, 1891; Sinai, Werner, 1893; Aden (*Yerbury*), Anderson, 1895; Hadramut, Anderson, 1896; Medina District, Hejaz, Anderson, 1896.

[1] Zool. Beitr. iii. 1798, p. 113.

[2] Prof. Boettger had not seen the specimen referred to this species, but recorded it on the verbal authority of Dr. Buck.

Agama pallida, Reuss, 1834.

Arabia, Sinaitic Peninsula, Reuss, 1834; Sinaitic Peninsula (*Hart*), Mt. Sinai, Boulenger, 1885; Sinai and Arabah, Hart, 1891; Aïn Musa and Tor, Sinaitic Peninsula, Anderson, 1896.

Agama ruderata, Olivier.

N. Arabia, Olivier; near Medina, Anderson, 1896.

Agama jayakari, Anderson, 1896.

Maskat (*Jayakar*), Anderson, 1896.

Agama flavimaculata, Rüppell, 1835.

Jiddah, Rüppell, 1835; Arabia, Boulenger, 1885; Hadramut, Anderson, 1896; Medina District, Hejaz, Anderson, 1896.

Agama cyanogaster, Rüppell [1], 1835.

Arabia (*Botta*), A. Duméril, 1851.

Agama adramitana, Anderson, 1896.

Hadramut, Anderson, 1896.

Agama stellio, Hasselq. & Linn., 1757.

Arabia (*Rüppell*), Heyden, 1827; Mt. Sinai, Boulenger, 1885; Tor, Sinaitic Peninsula, Anderson, 1896; Medina District, Hejaz, Anderson, 1896.

Phrynocephalus arabicus, Anderson, 1894.

Hadramut, Anderson, 1894.

Uromastix ornatus, Heyden, 1827.

Moïlab (*Rüppell*), Heyden, 1827; Mt. Sinai, Werner, 1893.

Uromastix ægyptius, Hasselquist & Linn., 1757.

Midian (*Sir R. Burton*), Günther, 1878; Maskat (*Jayakar*), Boulenger, 1887.

Uromastix hardwickii, Gray, 1827.

Makallah, Carter, 1864; Boulenger, 1885.

Uromastix (Aporoscelis) benti, Anderson, 1894.

Hadramut, Anderson, 1894; 50 miles from Aden (*Nurse*), Anderson, 1896.

Varanus griseus, Daudin, 1803.

Maskat (*Jayakar*), Boulenger, 1887; Aden (*Yerbury*), Anderson, 1895; Hadramut, Anderson, 1896.

Lacerta jayakari, Blgr., 1887.

Maskat (*Jayakar*), Boulenger, 1887.

---

[1] Duméril and Bibron erroneously give Jiddah as the locality whence the type of *A. cyanogaster* was obtained; and many authors have repeated their statement. It came from Massowah.

Latastia longicaudata, Reuss, 1834.

Tor, Sinaitic Peninsula (*Rüppell*), Reuss, 1834.

Latastia neumanni, Matschie, 1893.

Aden, Matschie, 1893; Anderson, 1895.

Acanthodactylus boskianus, Daud., 1803.

Midian (*Burton*), Günther, 1878; Sinaitic Peninsula (*Hart*), Mt. Sinai, Boulenger, 1887; Mt. Sinai and Arabah, Hart, 1891; Aden, Matschie, 1893; Sinai, Werner, 1893, (*Yerbury*) Anderson, 1895; Hadramut, Anderson, 1896.

Acanthodactylus scutellatus, Audouin, 1827.

Mount Sinai, Boulenger, 1887.

Acanthodactylus cantoris, Gthr., 1864.

Aden (*Yerbury*), Anderson, 1895; Hadramut, Anderson, 1896.

Eremias guttulata, Licht., 1823.

Arabia (*Westphal-Castlenau*), 1870; Sinaitic Peninsula (*Hart*), Mt. Sinai, Boulenger, 1887; Arabia and Sinai, Hart, 1891; Hadramut, Anderson, 1896.

Eremias brevirostris, Blanf., 1876.

Hadramut, Anderson, 1896.

Eremias rubropunctata, Licht., 1823.

Sinaitic Peninsula (*Hart*); Mt. Sinai, Boulenger, 1887, Hart, 1891.

Eremias mucronata, Blanf., 1870.

Mt. Sinai, Boulenger, 1887.

Mabuia brevicollis, Wiegm., 1837.

Aden (*Neumann*), Matschie, 1892, (*Yerbury*) Anderson, 1895; Hadramut, Anderson, 1896.

Mabuia tessellata, Anderson, 1895.

Aden (*Yerbury*), Anderson, 1895.

Mabuia septemtæniata, Reuss, 1834.

Arabia and Maskat (*Arnaud*), A. Duméril, 1851; Maskat (*Blanford*), Boulenger, 1887; Maskat, Werner, 1894.

Mabuia quinquetæniata, Licht., 1823.

Mount Sinai, Boulenger, 1887.

Ablepharus pannonicus, Fitz., 1829.

Arabia Petræa (*Rüppell*), Gray, 1839; Rüppell, 1845; Boettger, 1893.

Scincus hemprichii, Wiegm., 1837.

Arabia, Boettger, 1892; Aden (*Yerbury*), Anderson, 1895.

Scincus conirostris, Blanf., 1881.
Hadramut, Anderson, 1896.
Scincus mitranus, Anderson, 1871.
Arabia, Anderson, 1871.
Scincus meccensis, Wiegm., 1837.
Arabia, Wiegmann, 1837.
Scincus muscatensis, Murray, 1886.
Maskat, Murray, 1886; Maskat (*Jayakar*), Boulenger, 1887;
Island of Bahrein (*Bornmüller*), Werner, 1894.
Chalcides (Gongylus) ocellatus [1], Forskål, 1775.
Arabia (*Arnaud*), A. Duméril, 1851; Sinaitic Peninsula
(*Hart*), Boulenger, 1887 ; Maskat (*Jayakar*), 1887 ; Sinai
and Arabah, Hart, 1891; Aden, Boettger, 1892, Matschie,
1893 ; Maskat (*Bornmüller*), Werner, 1894; Aden (*Yerbury*), Anderson, 1895 ; Hadramut, Anderson, 1896.
Chalcides (Sphænops) sepoides, Audouin, 1827.
Tor, Sinaitic Peninsula, Peters, 1864; Maskat (*Jayakar*),
Boulenger, 1887 ; Sinaitic Peninsula (*Hart*), Boulenger,
1887 ; Wádí Ghurandel and Mt. Sinai, Hart, 1891.

RHIPTOGLOSSA.

Chamæleon vulgaris, Daud., 1803.
Arabia (*Christy*), Gray, 1864; Sinai and Aïn Musa, Hart,
1891 ; Aïn Musa, Anderson, 1896.
Chamæleon calcarifer, Peters, 1870.
Aden (*Yerbury*), Boulenger, 1887; (*Neumann*) Matschie,
1893 ; (*Yerbury*) Anderson, 1895; Hadramut, Anderson,
1896 ; (*Nurse*) Anderson, 1896.
Chamæleon calyptratus, A. Duméril, 1851.
Yemen (*Cairo Museum*), Anderson, 1896.

OPHIDIA.

Typhlops vermicularis, Merrem, 1820.
Sinai, Duméril & Bibron, 1844.
Glauconia nursii, Anders., 1896.
Aden (*Nurse*), Anderson, 1896.
Eryx jaculus, Hasselq. & Linn., 1757.
Arabia, Duméril & Bibron, 1844.

---

[1] Forskål gives Egypt as the habitat of the species and makes no mention
of Arabia, but some authors have quoted him as the authority of its occurrence
in Arabia.

Eryx jayakari, Blgr., 1888.

Maskat (*Jayakar*), Boulenger, 1893.

Zamenis rhodorhachis, Jan, 1865.

Midian (*Burton*), Günther, 1878; Maskat (*Jayakar*), Boulenger, 1887 [1]; Aden (*Schmidt*), Boettger, 1892 [2], (*Yerbury*) Boulenger, 1893, (*Yerbury*) Anderson, 1895; Hadramut, Anderson, 1896.

Zamenis karelinii, Brandt, 1838.

Ras Massendam at entrance to Persian Gulf, Bedriaga, 1879.

Zamenis elegantissimus, Günther, 1878.

Mountain east of El Muwáylah, Midian (*Burton*), Günther, 1878; Akabah, Hart, 1891.

Zamenis diadema, Schlegel, 1837.

Sandy Coast region of the Tehama, Midian (*Burton*), Günther, 1878; Maskat (*Jayakar*), Boulenger, 1887; Mount Hor, Arabia Petræa, Hart, 1891; Maskat (*Jayakar*), Boulenger, 1893; Hadramut, Anderson, 1896.

Lytorhynchus diadema, Dum. & Bibr., 1854.

Maskat (*Jayakar*), Boulenger, 1887; Aden (*Neumann*), Matschie, 1893.

Oligodon melanocephalus, Jan, 1862.

Arabia Petræa, Hart, 1891.

Tarbophis dhara, Forskål, 1775.

Yemen, Forskål, 1775; Medina District, Hejaz, Anderson 1896.

Tarbophis guentheri, Anderson, 1895.

Maskat (*Jayakar*), Boulenger, 1887 [3]; Aden (*Yerbury*), Anderson; Hadramut, Anderson, 1896.

Cœlopeltis monspessulana, Hermann, 1804.

Mount Sinai, Hart, 1891; Sinai, Werner, 1893.

Cœlopeltis moilensis, Reuss, 1834.

Moilah, Midian (*Rüppell*), Reuss, 1834; Aden (*Yerbury*), Anderson, 1895; Hadramut, Anderson, 1896.

Psammophis schokari, Forskål, 1775.

Yemen, "frequens in sylvis montosis," Forskål, 1775; Tor, Sinaitic Peninsula (*Rüppell*), Reuss, 1834; Maskat

[1] Recorded as *Z. ventrimaculatus*, Gray.
[2] Recorded by Prof. Boettger as *Z. ladacensis*, Anders.
[3] Recorded as *Dipsas obtusa*, Reuss.

(*Jayakar*), Boulenger, 1878 [1]; Aden (*Yerbury*), Anderson, 1895; Aïn Musa, Suez, Anderson, 1896; Hadramut, Anderson, 1896.

Psammophis punctulatus, Dum. & Bibr., 1854.
Arabia (*Arnaud*), Duméril & Bibron, 1854.

Vipera arietans, Merrem, 1820.
Hadramut, Anderson, 1896.

Cerastes cornutus, Hasselq. & Linn., 1757.
Arabah, Arabia Petræa, Strauch, 1862; Sinai, Werner, 1893.

Echis carinatus, Schn., 1801.   '
Sandy coast of the Tehama, Midian (*Burton*), Günther, 1878; Maskat (*Jayakar*), Boulenger, 1887; Aden (*Schmidt*), Boettger, 1892; Hadramut, Anderson, 1896.

Echis coloratus, Gthr., 1878.
Jebel Shárr, 4500 ft., Midian (*Burton*), Günther, 1878; Maskat (*Jayakar*), Boulenger, 1887; Hadramut, Anderson, 1896.

BATRACHIA.

ECAUDATA.

Rana esculenta, Linn., 1766.
Ghor, South end of Dead Sea, Hart, 1891.

Rana cyanophlyctis, Schn., 1799.
Arabia, Peters, 1863, Boulenger, 1882; Aden (*Neumann*), Matschie, 1893, (*Yerbury*) Anderson, 1895; Hadramut, Anderson, 1896.

Bufo viridis, Laur., 1768.
Arabia Petræa (*Rüppell*), Heyden, 1827; Arabia (*Burton*), Boulenger, 1880 & 1882.

Bufo andersoni, Blgr., 1883.
Maskat (*Jayakar*), Boulenger, 1887, Werner, 1894; Aden (*Yerbury*), Anderson, 1895.

Bufo pentoni, Anderson, 1893.
Aden (*Neumann*), Matschie, 1893; (*Yerbury*) Anderson, 1895.

Bufo regularis, Reuss, 1834.
Midian (*Burton*), Boulenger, 1882.

[1] Recorded as *P. leithii,* Gthr.

GENERAL DISTRIBUTION OF THE SPECIES.

	AFRICA.	ASIA.	EUROPE.
Testudo elegans, *Schoepff*.	....	Ceylon, India, Sind, E. Arabia.	
Testudo leithii, *Gthr.* ...	Lower Egypt.	Between Ismailia and El Arisch ; Lower Syria.	
Stenodactylus elegans, *Fitz.*	Algerian Sahara to Egypt, Nubia, Southern Abyssinia and Eastern Sudan ; Lake Rudolph.	N.W. Arabia, S. Syria, Western Bejudab desert.	? Island of Syra.
Stenodactylus (C.) pulcher, *Anders.*	....	S.E. Arabia.	
Stenodactylus (C.) doriæ, *Blanf.*	....	Persia, Arabia.	
Bunopus tuberculatus, *Blanf.*	....	Sind, Afghanistan, Baluchistan, South-Eastern Persia, E. Arabia.	
Bunopus blanfordii, *Strauch.*	Egypt.	S. E. Arabia.	
Gymnodactylus scaber, *Heyden.*	Egypt, Abyssinia.	Arabia, S. Mesopotamia, Persia, Afghanistan, Sind.	
Pristurus flavipunctatus, *Rüppell.*	Western Somaliland, Abyssinia.	S. Arabia.	
Pristurus rupestris, *Blanf.*	....	Sind, S. Persia, S.E. Arabia.	
Pristurus crucifer, *Val.* .	Abyssinia, Somaliland.	S. Arabia.	
Pristurus collaris, *Steind.*	....	S.E. Arabia.	
Pristurus carteri, *Gray* .	....	S.E. Arabia.	
Ptyodactylus hasselquistii, *Donndorf.*	Algeria to Egypt, coast of Red Sea, Shoa to near Assab.	Syria, W., N., & E. Arabia.	
Hemidactylus sinaitus, *Blgr.*	West Coast of Red Sea.	Arabia.	
Hemidactylus turcicus, *Linn.*	Algeria to Egypt, Eastern Sudan, Senaar, coast of Red Sea, Abyssinia.	N. & S.E. Arabia, Syria, Asia Minor, Cyprus, Persia, Sind.	Borders of Mediterranean.
Hemidactylus yerburii, *Anders.*	....	Arabia.	
Hemidactylus flaviviridis, *Rüppell.*	Coast of Red Sea.	Arabia, Persia, Baluchistan, Afghanistan, Indo-Malaya.	
Tarentola annularis, *Is. Geoffr.*	Egypt, Nubia, E. Sudan, S. Abyssinia.	N.W. Arabia.	
Tarentola mauritanica, *Linn.*	Algeria to Egypt.	N.W. Arabia, Cyprus.	Borders of Mediterranean.
Agama sinaita, *Heyden* .	Egypt, ? Sennaar.	N., S., & E. Arabia, Syria.	
Agama pallida, *Reuss* ...	Egypt.	N.W. Arabia.	
Agama ruderata, *Olivier*	....	N. & W. Arabia, Syria to Sind.	
Agama jayakari, *Anders.*	....	E. Arabia.	

	AFRICA.	ASIA.	EUROPE.
Agama flavimaculata, *Rüppell.*	....	W. Arabia.	
Agama cyanogaster, *Rüppell.*	Abyssinia, Somaliland.	W. Arabia.	
Agama adramitana, *Anders.*	....	S.E. Arabia.	
Agama stellio, *Hasselq. & Linn.*	Egypt.	S. Caucasus, Asia Minor, Cyprus, Syria, N. & W. Arabia.	S.E. Europe.
Phrynocephalus arabicus, *Anders.*	....	S.E. Arabia.	
Uromastix ornatus, *Heyden.*	....	W. Arabia.	
Uromastix ægyptius, *Hasselq. & Linn.*	Algerian Sahara, Egypt.	W. & E. Arabia.	
Uromastix hardwickii, *Gray.*	....	N.W. India, Baluchistan, S.E. Arabia.	
Uromastix (A.) benti, *Anders.*	....	S.E. & S. Arabia.	
Varanus griseus, *Daud.*	North Sahara, Algeria, to Egypt, Eastern Sudan.	S., S.E., & E. Arabia, S. Syria, Persia, Caspian Province, Afghanistan, N.W. India.	
Lacerta jayakari, *Blgr.*	....	E. Arabia.	
Latastia longicaudata. *Reuss.*	E. Sudan, Abyssinia, Somaliland, Taita.	N.W. Arabia.	
Latastia neumanni, *Matschie.*	....	S. Arabia.	
Acanthodactylus boskianus, *Daud.*	Algeria to Egypt, Nile Valley, Abyssinia, Eastern Sudan.	Southern Syria, N., S., & S.E. Arabia.	
Acanthodactylus scutellatus, *Aud.*	Senegambia, Atlantic (Cape Jubi) to Red Sea, Nile Valley (desert), Abyssinia, Somaliland.	Southern Syria, N. Arabia.	
Acanthodactylus cantoris, *Gthr.*	....	S. & E. Arabia, S. Persia, Baluchistan, Afghanistan, Sind, N.W. India.	
Eremias guttulata, *Licht.*	Mogador to Red Sea, Nile Valley (desert); Socotra.	S. Syria, N. Arabia, Syria, Persia, Baluchistan, Sind.	
Eremias brevirostris, *Blanf.*	....	Persia, Bekâ'a (Coelesyria), S.E. Arabia.	
Eremias rubropunctata, *Licht.*	Egypt.	N.W. Arabia.	
Eremias mucronata, *Blanf.*	Eastern Sudan, Abyssinia, Somaliland, Berbera, Tana river.	N.W. Arabia.	
Mabuia brevicollis, *Wiegm.*	Abyssinia.	S. Arabia.	
Mabuia tessellata, *Anders.*	....	S. Arabia.	
Mabuia septemtæniata, *Reuss.*	Abyssinia.	E. Arabia, Syria, Asia Minor, Persia, Sind.	

	AFRICA.	ASIA.	EUROPE.
Mabuia quinquetæniata, *Licht.*	Senegal, Benguela, Mozambique,to Delta of Nile, Abyssinia, Easteru Sudan.	N.W. Arabia.	
Ablepharus pannonicus, *Fitz.*	....	N.W. Arabia, Syria, Asia Minor, Cyprus, Island of Syra.	Hungary, Roumelia, Albania, Greece.
Scincus hemprichii, *Wiegm.*	Abyssinia.	S. Arabia.	
Scincuscouirostris,*Blanf.*	....	S. Persia, S.E. Arabia.	
Scincus mitranus, *Anders.*	....	Arabia.	
Scincusmeccensis,*Wiegm.*	....	W. Arabia.	
Scincus muscatensis, *Murray.*	....	E. Arabia.	
Chalcides (G.) ocellatus, *Forskål.*	El Gada (Cape Jubl) to Egypt, NileValley, coast of Red Sea, Abyssinia, Berbera, Somaliland.	Sind, Persia, Arabia, Syria, Chios, Rhodes, Cyprus.	Sardinia, Sicily, Malta, Lampedusa, Crete.
Chalcides (S.) sepoides, *Aud.*	Senegambia, El Gada (Cape Jubi) to Egypt, Nile Valley to Wadi Halfa, Somaliland.	Syria, N.W. and E. Arabia.	
Chamæleon vulgaris, *Daud.*	Mogador to Egypt.	N.W. Arabia, Syria, Cyprus, Asia Minor, Chios.	S. Spain.
Chamæloon calcarifer, *Peters.*	....	S. and S.E. Arabia.	
Chamæleon calyptratus, *A. Duméril.*	Nile Valley.	S.W. Arabia.	
Typhlops vermicularis, *Merr.*	....	Caucasus, Turkestan, Afghanistan, E. Persia, Asia Minor,Syria, Palestine, N.W. Arabia.	Turkey, Greece, Ionian Islands, Rhodes, Cyprus.
Glauconia nursii, *Anders.*	....	S.E. Arabia.	
Eryx jaculus, *Hasselq. & Linn.*	Algeria to Lower Egypt.	N.W. Arabia, Syria, Asia Minor, Persia, Turkestan, Afghanistan.	Greece and its islands, Corfu.
Eryx jayakari, *Blgr.* ...	....	E. Arabia.	
Zamenis rhodorhachis, *Jan.*	Egypt, Somaliland.	Arabia, Syria, Persia, Baluchistan, Punjab, Western Himalayas.	
Zamenis karelinii,*Brandt.*	....	Turkestan, Afghanistan, Baluchistan, Persia, E. Arabia.	
Zamenis elegantissimus, *Gthr.*	....	N.W. Arabia.	
Zamenis diadema, *Schleg.*	North Sahara, Algeria to Egypt, Eastern Sudan.	Arabia, Syria, Asia Minor, Persia, Baluchistan, Afghanistan, Turkestan, Sind, Western Himalayas, Punjab, N.W. Provinces India and Western Konkan (Bombay).	

	AFRICA.	ASIA.	EUROPE.
Lytorhynchus diadema, *Dum. & Bibr.*	Algerian Sahara to Egypt, Senaar.	E. & S. Arabia, Syria, Persia.	
Oligodon melanocephalus, *Jan.*	Lower Egypt (Cairo).	N. Arabia, Syria.	
Tarbophis dhara, *Forskål.*	Nile Valley.	W. Arabia.	
Tarbophis guentheri, *Anders.*	....	S. & S.E. Arabia.	
Cœlopeltis monapessulana, *Hermann.*	Marocco to Egypt.	N.W.Arabia,Syria,Asia Minor,Cyprus, Lower Caucasus, Persia.	Southern Europe.
Cœlopeltis moilensis, *Reuss.*	Algerian Sahara to Egypt, Nubia.	W. & S. Arabia, Western Persia.	
Psammophis schokari, *Forskål.*	Marocco, Algerian Sahara to Egypt, Nile Valley to Khartoum, Eastern Sudan.	Arabia, Syria, Persia, Baluchistan, Afghanistan, Sind.	
Psammophis punctulatus, *Dum. & Bibr.*	Somaliland, Mozambique.	W. Arabia.	
Vipera arietans, *Merr.* ...	Cape of Good Hope to Senegambia and Marocco, and to Kordofan, Abyssinia.	S.E. Arabia.	
Cerastes cornutus, *Hasselq. & Linn.*	Algerian Sahara to coast of Red Sea, Nile Valley to Nubia.	N.W. Arabia, S. Palestine (desert).	
Echis carinatus, *Schn.* ...	W. Africa, Northern Sahara to coast of Red Sea, Nile Valley (desert), Eastern Sudan, Abyssinia, Somaliland.	Arabia, desert S. Palestine, Persia, Baluchistan, Afghanistan, Turkestan, Sind, India.	
Echis coloratus, *Gthr.* ...	Island of Socotra.	W. Arabia, S. end of Dead Sea, Jericho.	
Rana esculents, *Linn.* ...	N. Africa.	Palæarctic region, N.W. Arabia.	Palæarctic region.
Rana oyanophlyctis, *Schn.*	....	Malay Peninsula to Baluchistan; Ceylon; S.E. Arabia.	
Bufo viridis, *Laur.*	N. Africa.	CentralAsia and southwards to the Himalayas ; N.W. Arabia.	East of Rhine & Rhone.
Bufo andersoni, *Blgr.* ...	....	Agra District, Rajputana, Sind ; S.E. Arabia.	
Bufo pentoni, *Anders.* ..	Suakin.	S. Arabia.	
Bufo regularis, *Reuss.* ...	N.E. Africa.	N.W. Arabia.	

Species essentially Arabian.

Stenodactylus (C.) pulcher, Anders.
Pristurus collaris, Steindachn.
    ,,   *carteri*, Gray.
Hemidactylus yerburii, Anders.
Agama jayakari, Anders.
    ,,   *flavimaculata*, Rüppell.
    ,,   *adramitana*, Anders.
Phrynocephalus arabicus, Anders.
Uromastix ornatus, Heyden.
    ,,   *(Aporoscelis) benti*, Anders.
Lacerta jayakari, Blgr.
Latastia neumanni, Matschie.
Mabuia tessellata, Anders.
Scincus mitranus, Anders.
    ,,   *meccensis*, Wiegm.
    ,,   *muscatensis*, Murray.
Chamæleon calcarifer, Peters.
Glauconia nursii, Anders.
Eryx jayakari, Blgr.
Zamenis elegantissimus, Gthr.
Tarbophis guentheri, Anders.

Western Arabia, Palestine, Syria, and Socotra.

Echis coloratus, Gthr.

Species confined to Africa and Arabia.

Bunopus blanfordii, Strauch.
Pristurus flavipunctatus, Rüppell.
    ,,   *crucifer*, Val.
Hemidactylus sinaitus, Blgr.
Tarentola annularis, Is. Geoffr.
Agama pallida, Reuss.
    ,,   *cyanogaster*, Rüppell.
Uromastix ægyptius, Hasselq. & Linn.
Latastia longicaudata, Reuss.
Eremias rubropunctata, Licht.

Eremias mucronata, Blanf.
Mabuia brevicollis, Wiegm.
„   *quinquetæniata*, Licht.
Scincus hemprichii, Wiegm.
Chamæleon calyptratus, A. Duméril.
Tarbophis dhara, Forskål.
Psammophis punctulatus, Dum. & Bibr.
Vipera arietans, Merr.

Species found in Africa and Arabia, but extending to the North,
or to the North-East beyond Arabia.

Testudo leithii, Gthr.
Stenodactylus elegans, Fitz.
Gymnodactylus scaber, Heyden.
Ptyodactylus hasselquistii, Donndorf.
Agama sinaita, Heyden.
Varanus griseus, Daud.
Acanthodactylus boskianus, Daud.
„   *scutellatus*, Aud.
Eremias guttulata, Licht.
Mabuia septemtæniata, Reuss.
Chalcides (S.) sepoides, Aud.
Zamenis rhodorachis, Jan.
Lytorhynchus diadema, Dum. & Bibr.
Cœlopeltis moilensis, Reuss.
Psammophis schokari, Forskål.
Cerastes cornutus, Hasselq. & Linn.

Asiatic species not extending to the West of the Persian Gulf
and not entering Africa.

Testudo elegans, Schoepff.
Bunopus tuberculatus, Blanf.
Pristurus rupestris, Blanf.
Uromastix hardwickii, Gray.
Acanthodactylus cantoris, Gthr.
Eremias brevirostris, Blanf.

Southern Asiatic to African Coast of Red Sea.

Hemidactylus flaviviridis, Rüppell.

Turkestan and Persian species not extending to West of the
Persian Gulf, and not entering Africa.
Stenodactylus (C.) doriæ, Blanf.
Scincus conirostris, Blanf.
Zamenis karelinii, Brandt.

Asiatico-African species.

Zamenis diadema, Schlegel.
Echis carinatus, Schn.

South-Eastern European and South-Western Asiatic species
not entering Africa.
Ablepharus pannonicus, Fitz.
Typhlops vermicularis, Merr.

South-Western Asiatic species not entering Africa.
Agama ruderata, Olivier.

South-Eastern European and South-Western Asiatic species
entering Egypt.
Agama stellio, Hasselq. & Linn.

Central and South-Western Asiatic and South-Eastern
European species extending to North Africa.
Eryx jaculus, Hasselq. & Linn.

South-Western Asiatic species extending to Egypt.
Oligodon melanocephalus, Jan.

European side of Mediterranean, South-Western Asia
(Sind to Syria), and North-Eastern Africa.
Hemidactylus turcicus, Linn.

South-Western and Southern Europe and North Africa.
Tarentola mauritanica, Linn.

Sardinia, Sicily, Greece, Cyprus, South-Western Asia (Sind
to Syria), North and North-Eastern Africa.
Chalcides (G.) ocellatus, Forskål.

South-Western Europe (Spain) and North Africa.
Chamæleon vulgaris, Daud.

South and South-Eastern Europe, South-Western Asia
(Persia to Syria), Cyprus, and North Africa.
Cœlopeltis monspessulana, Hermann.

Palæarctic Region.
Rana esculenta, Linn.

Europe east of the Rhone and Rhine, North Africa, Central
Asia southwards to the Himalayas.
Bufo viridis, Laur.

African and Arabian species.
Bufo regularis, Reuss.
„ *pentoni*, Anders.

Malayan Peninsula to Baluchistan, Ceylon.
Rana cyanophlyctis, Schn.

Agra District to Sind.
Bufo andersoni, Blgr.

LITERATURE.

1751. Hasselquist, Acta Soc. Reg. Sc. Upsal.

1757. Hasselquist's 'Iter Palæstinum' was edited by Linnæus, who took the responsibility for the names of the species enumerated.

1766. Linnæus, Syst. Nat. i., xii. ed.

1768. Laurenti, Syn. Rept.

1775. Forskål, Descr. Anim. etc.

1792–98. Donndorf, Zool. Beytr.

1792–1801. Schoepff, Hist. Test.

1799–1801. Schneider, Hist. Amph.

1801–7. Olivier, Voy. Emp. Othoman.

1802. Daudin, Hist. Rept. iii.

1803. Daudin, Hist. Rept. iv. & v.

1820. Merrem, Syst. Amph.

1823. Lichtenstein, Verz. Doubl. Mus. Berl.

1827. Geoffroy, Is., & Audouin, Descr. de l'Égypte, Reptiles[1].

1829. Cuvier, Règne Anim.

1829. Fitzinger, Verh. Ges. Nat. Fr. Berlin, i.

1834. Reuss, Mus. Senck. i.

1835. Rüppell, N. Wirbelth. F. Abyss., Amphibien.

1837. Duméril & Bibron, Erpét. Génl. iv.

1837. Schlegel, Phys. Serp.

1837. Wiegmann, Arch. Naturg.

1839. Gray, Ann. & Mag. N. H. ii.

1843–45. Rüppell, Mus. Senck. iii.

1844. Duméril & Bibron, Erpét. Génl. vi.

1851. Duméril, Cat. Méth. Rept.

1854. Duméril & Bibron, Erpét. Génl. vii.

1861. Valenciennes, Compte-Rend. liii.

1861–81. Jan, Icon. Gén.

1862. Jan, Arch. Zool. Anat. Phys. ii.

1862. Strauch, Mém. Ac. St. Pétersb. (vii.) iv.

1863. Jan, Elenco Sist. Ofidi.

---

[1] Is. Geoffroy St. Hilaire, in his preface (p. 2) to the fifth volume of the work entitled 'Voyage autour du Monde sur la Frégate La Venus,' gives 1827 as the date of his sections of the above work. On the other hand, Duméril in the third volume of the Erpét. Génl. p. 284, mentions 1828 as the date of its publication.

1863–64. Gray, Proc. Zool. Soc.

1864. Carter, Proc. Zool. Soc.

1864. Günther, Rept. Brit. Ind.

1864. Peters, Mon. Berl. Ac.

1867. Steindachner, Reise Freg. 'Novara,' Rept.

1869. Peters, Mon. Berl. Ac.

1869. Strauch, Mém. Ac. St. Pétersb. (vii.) xiv. no. 6.

1870–71. Peters, Mon. Berl. Ac.

1874. Blanford, Ann. & Mag. N. H. (4) xii.

1877. Boettger, Zeitschr. ges. Naturw. Berl. (Giebel) (2) i.

1878. Günther, Proc. Zool. Soc. Burton's 'Gold Mines of Midian.'

1878–80. Boettger, Ber. Senck. Nat. Ges.

1879. Bedriaga, Bull. Soc. Nat. Moscou.

1880. Boulenger, Proc. Zool. Soc.

1882. Boulenger, Cat. Batr. Sal. B. M.

1882. Boulenger, Cat. Batr. Grad. B. M.

1882. Vaillant, Miss. Révoil, Rept. et Batr.

1883. Boulenger, Ann. Mag. N. H. (5) xii.

1883. Lortet, Arch. Mus. Lyon. iii.

1884. Murray, Vert. Zool. Sind.

1885. Boulenger, Cat. Liz. B. M. i.

1886. Murray, Ann. & Mag. N. H. (5) xvii.

1887. Strauch, Mém. Ac. St. Pétersb. (vii.) xxxv. no. 2.

1887. Boulenger, Cat. Liz. B. M. iii.

1887. Boulenger, Aun. & Mag. N. H. (5) xx.

1888. Boulenger, Ann. & Mag. N. H. (6) ii.

1889. Boulenger, Cat. Chelonians B. M.

1891. Boulenger, Trans. Zool. Soc. xiii.; Ann. Mus. Genova, (2) xii., xxxii.

1891. Hart, Fauna and Flora of Sinai, Petra, &c.

1892. Boettger, Ber. Offenb. Ver. Nat.

1893. Boulenger, Cat. Snakes B. M. i.

1893. Boettger, Kat. Rept. Samml. Mus. Senck.

1893. Matschie, Sitz. Ges. Naturf. Fr. Berl.

1893. Werner, Verh. zool.-bot. Ges. Wien.

1894. Paracca, Boll. Mus. Torino.

1894. Anderson, Ann. & Mag. N. H. (6) xiv.

1894. Boulenger, Cat. Snakes B. M. ii.

1894. Werner, Verh. zool.-bot. Ges. Wien.

1895. Anderson, Proc. Zool. Soc.

PART VI.

A PRELIMINARY LIST OF THE

REPTILIA AND BATRACHIA OF EGYPT

(FROM THE DELTA TO WÁDÍ HALFA)

AND OF THE DISTRICT OF SUAKIN.

I AVAIL myself of this opportunity to publish this simple Name-list, as it may be useful for comparison with the Arabian List; and, moreover, it affords me the means of expressing my thanks to all those who have been so good as to assist me in procuring specimens for the furtherance of my work on these sections of the Fauna of Egypt[1].

A complete set of all the species represented in the Collection will be presented to the British Museum, and another set, as perfect as it may be possible to make it, will also be presented to the Museum of the Medical School of Cairo. The remaining duplicates will be placed at the disposal of the former Institution, for exchange with other Museums.

I append to the List the names of certain species not represented in the Collection, but which have been recorded from Egypt; so that it may be seen at a glance what is the present state of our knowledge regarding these sections of the fauna in the areas indicated, and what are the desiderata necessary to make the Collection complete. The mention of these desiderata may, I hope, possibly lead to their being procured by some of those who have so kindly assisted me in the past.

The Collection comprises 982 specimens of Lizards, 233 of Snakes, and 89 of Batrachia. To these have to be added 1 Crocodile (juv.), 3 land Tortoises, and 3 river Turtles.

[1] I have included the Suez district to Aïn Musa.

REPTILIA.

EMYDOSAURIA.

CROCODILUS NILOTICUS, Daud.

1 juv. Wádí Halfa. Surgeon-Captain R. H. Penton.

CHELONIA.

TESTUDO LEITHII, Günther.

1 ♂ and 2 ♀. Neighbourhood of Alexandria.

TRIONYX TRIUNGUIS, Forskål.

1 ♀. Nile at Cairo. Dr. Walter Innes.
1 juv. ♀. Above First Cataract of Nile.
1 juv. ♀. Wádí Halfa. Major Henry d'Alton Harkness.

SQUAMATA.

LACERTILIA.

STENODACTYLUS ELEGANS, Fitzinger.

Ascalabotes sthenodactylus, Licht. Verz. Doubl. Berl. Mus.
1823, p. 102.
Stenodactylus elegans, Fitz. N. Class. Rept. 1826, p. 47.
Agame ponctué, Is. Geoffr. Descr. de l'Égypte, Nat. Hist. i.
(1827) pp. 129–130, pl. v. fig. 2.
Stenodactylus guttatus, Cuv. Règn. An. nouv. ed. 2, ii. 1829,
p. 58; Dum. & Bibr. part iii. 1836, p. 434, *sed non* pl. 34, no. 2.
Trapelus savignyi, Aud. Descr. de l'Égypte, Nat. Hist. i. 1827,
p. 167; Suppl. pl. i. figs. 3.1, 3.2, 3.3; Gasco, Viaggio in
Egitto, pt. ii. 1876, p. 115.
Tolarenta wilkinsonii, Gray, Zool. Misc. 1831, p. 58.
Stenodactylus mauritanicus, Guich. Explor. Algér., Zool. v.
1850, p. 5, pl. i. fig. 1.

1 ♂. Mandara, east of Alexandria. Dr. Walter Innes.
1 ♂ and 1 ♀. Ramleh, east of Alexandria.
8 ♂ and 12 ♀. From around pyramids of Gizeh.
3 ♀. Tel el Amarna. Professor W. M. Flinders Petrie,
D.C.L., &c.

3 ♂ and 1 ♀. Luxor Desert.

1 ♂. Wádí Halfa. Surgeon-Captain R. H. Penton.

1 ♂ and 1 ♀. Suakin.

2 ♀. Durrur, about 60 miles N. of Suakin.

1 ♂. Ras Gharíb, Gulf of Suez. Mr. James Robertson.

STENODACTYLUS PETRII, n. sp.

Trapelus savignyi, Aud. var. (*non* Dum. & Bibr.) *op. cit.*
p. 168; Suppl. Rept. pl. i. figs. 4. 1, 4. 2, and 4. 3, *sed non*
Suppl. Rept. pl. i. figs. 3. 1, 3. 2, and 3. 3.

Stenodactylus guttatus, Dum. & Bibr. part iii. (1836), p. 434,
pl. 34, no. 2.

1 ♂ and 2 ♀. Tel el Amarna. Professor W. M. Flinders
Petrie, D.C.L.

Head large, very distinct from the neck; cheeks swollen;
snout short and moderately pointed; nostril swollen, defined by
the first labial and three nasals; 12 to 15 upper and 11 to 14
lower labials; mental large, as broad as the rostral and first
labial. Eye very large; ear small, slightly oval and vertical.
Body covered with small smooth, slightly convex scales, generally
longer than broad, polygonal and somewhat smaller on the middle
of the back than on the sides, where they are more rounded;
they are largest on the snout, hexagonal, very slightly convex
and rugose; scales on the limbs as large as those on the sides,
tending to become imbricate and feebly keeled on the humeral
and femoral regions; scales on the under surface of the body
slightly imbricate, obscurely obtusely keeled. Under surface of
the digits with a longitudinal row of transverse tricarinate
lamellæ, more or less imbricate, with two rows of small, distinctly
pointed scales external to it; seven rows of scales on the upper
surface of the third toe, about its middle; the outer row of
dorsal scales of each digit forms a well-marked fringe most
strongly developed on the hind foot, each scale being antero-
posteriorly expanded at its base and curved distally into a sharp
point, the entire fringe being slightly downwardly curved. Scales
on the tail arranged in rings, largest on the upper surface, somewhat larger than the largest body-scales, longer than broad, and
more or less strongly keeled; those of the under surface much
smaller and rounded. Limbs long and slender; fore limb when
laid forwards reaches beyond the snout, and the hind limb in

advance of the shoulder. Digits moderately long and slender. Tail contracted behind the basal swelling, rapidly tapered to a fine point, shorter than the body and head. No præanal pores, and no enlarged scales on the position occupied by these structures. General colour of the upper parts pale but rich fawn, with irregular dark brown markings, most pronounced on the head, feeble on the upper surface of the trunk, and tending to anastomose; the most pronounced head-marking occurs behind the eye, and curving inwards tends to unite with its fellow of the opposite side; an ill-defined pale brown band from the ear along the side; tail banded to its tip with dark brown; chin to vent, and sides of belly whitish; under surface of limbs and tail yellowish.

	♀	♂
Snout to vent	60	54
Length of head	18	18
Width of head	14	14
Vent to tip of tail	53	51

This species is distinguished from *S. elegans*, Fitz., by its longer and broader head, by an additional row of scales on each side of the central digital lamellæ, by its tapered and finely pointed tail, and by its different coloration.

TROPIOCOLOTES TRIPOLITANUS, Peters.

Tropiocolotes tripolitanus, Peters, Mon. Berl. Ac. 1880, p. 306, pl. —. fig. 1; Blgr. Trans. Linn. Soc. xiii. 1891, p. 108.

Stenodactylus tripolitanus, Blgr. Cat. Liz. B. M. i. 1885, p. 19.

6 specimens from around pyramids of Gizeh, under stones.
Heretofore, known only from Tripoli and Tunisia.

TROPIOCOLOTES STEUDNERI, Peters.

Gymnodactylus steudneri, Peters, Mon. Berl. Ac. 1869, p. 788; Gasco, Viaggio in Egitto, pt. ii. 1876, p. 113 : Blgr. Cat. Liz. B. M. i. 1885, p. 34.

Stenodactylus petersii, Blgr. *op. cit.* i. p. 18, pl. iii. fig. 4.

Stenodactylus steudneri, Blgr. *op. cit.* iii. 1887, p. 480.

Tropiocolotes steudneri, Blgr. Trans. Zool. Soc. xiii. 1891, p. 108.

1. Neighbourhood of pyramids of Gizeh, under stones.
15. Margin of desert, Luxor ; dug out of small holes.
1. Desert of Philæ.

PTYODACTYLUS HASSELQUISTII, Donndorf.
1 ♂ and 1 ♀. Plain of Suez. = *P. guttatus*, Heyden.

Var. *siphonorhina*.
1 ♂. Abu Roash, near Gizeh. The late V. Ball, Esq., C.B.
1 ♂ and 1 ♀. Beni Hassan. M. W. Blackden, Esq.

Typical form.
1 ♀. Mokattam Hills, Cairo. Dr. Walter Innes.
1 ♂ and 1 ♀. Luxor.
2 ♀. In the dark recesses of a chamber in the temple of Medinet Habu.
1 ♀. Temple of Edfu.
2 ♂ and 2 ♀. Houses, Assuan.
5 ♂ and 2 ♀. Temple of Philæ.
1 ♀. Wádí Halfa.
3 ♂ and 3 ♀. Wádí Halfa. O. Charlton, Esq.

PRISTURUS FLAVIPUNCTATUS, Rüppell.
1 ♂. Suakin. Surgeon-Captain R. H. Penton.
3 ♂ and 5 ♀. Durrur.

HEMIDACTYLUS TURCICUS, Linn.
2 ♀. Maryut District, West of Alexandria.
2 ♀. Houses, Alexandria.
1 ♀. Mokattam Hills. Dr. Walter Innes.
1 ♀. Edfu, Upper Egypt.
2 ♂ and 3 ♀. Suakin. Surgeon-Captain R. H. Penton.
1 ♂ and 1 ♀. Suakin.
2 ♀ and 1 ♂. Island of Shadwan, Gulf of Suez. Mr. John Strathearn.
3 ♀. Ras Gharíb. Mr. James Robertson.
1 ♂. Shaluf, Suez.

HEMIDACTYLUS SINAITUS, Blgr.
1 juv. Wádí Halfa. Major Henry d'Alton Harkness.
4 ♂, 3 ♀, and 1 juv. Suakin. Surgeon-Captain R. H. Penton.
5 ♂ and 4 ♀. Suakin.
4 ♂ and 4 ♀. Durrur.

HEMIDACTYLUS FLAVIVIRIDIS, Rüppell.
1. Suez. Rev. Walter Statham.
1. Suakin. Surgeon-Captain R. H. Penton.
2. Suakin, British Officers' Mess House.

TARENTOLA ANNULARIS, Is. Geoffr.

1. A house, Cairo. Dr. Walter Innes.
2. Pyramids of Gizeh.
2. Mariette Bey's house, Sakhâra.
2. Minia. Major R. H. Brown, R.E.
1. Tel el Amarna. Professor W. M. Flinders Petrie, D.C.L.
1. Luxor.
2. Colossi of Memnon, Thebes.
1. Rocks, banks of Nile, Assuan.
2. Rocks, banks, of Nile, above First Cataract.
1. Wádí Halfa. Surgeon-Captain R. H. Penton.
2. Suakin. Colonel Sir Charles Holled Smith, C.B., K.C.M.G.
4. Houses, Suakin. Henry Barnham, Esq., H.B.M. Consul, Suakin.
1. Rocks of Dehilba, Suakin plain.
3. Erkowit, near Suakin.
6. Houses, Suakin.
1. Durrur.

TARENTOLA MAURITANICA Linn.

1. Cairo, houses. Dr. Walter Innes.
7. Abukir, on the walls of old windmills.
1. Mandara, east of Alexandria. Dr. Walter Innes.
2. Ramleh, east of Alexandria.
7. El Khreit, to the west of Lake Mareotis.

TARENTOLA EPHIPPIATA, O'Shaughnessy.

Tarentola ephippiata, O'Shaughn. Ann. Mag. N. H. (4) xvi.
1875, p. 264; Blgr. Ann. Mag. N. H. (6) xvi. 1895, p. 166.

1. Durrur.

Hitherto only recorded from West Africa and Somaliland.

AGAMA SINAITA, Heyden.

1 ♀. Plain of Suez.
1 ♂, 1 ♀, and 1 juv. Stony desert above Wádí Hoaf, Heluan.

AGAMA PALLIDA, Reuss.

Agama ruderata (non Oliv.), Aud. *op. cit.* p. 169, Suppl. Rept.
pl. i. fig. 6.
Agama pallida, Reuss, Mus. Senckenb. i. 1834, p. 38, pl. iii.
fig. 3.
Agama loricata, Reuss, *op. cit.* p. 40.

Agama nigrofasciata, Reuss, *op. cit.* p. 42.

Agama leucostygma, Reuss, *op. cit.* p. 44.

5 ♂ and 3 ♀.   Walls of houses, Suez.

3 ♂ and 2 ♀.   Between Ismailia and Suez.

2 ♂ and 2 ♀.   Beltim, between Rosetta and Damietta.   Dr. J. G. Rogers.

2 ♂ and 2 ♀.   Plain of Kafr Gamus, Matariyeh.

1 ♂ and 1 ♀.   Abbasiyeh.   Colonel H. M. L. Rundle, D.S.O.

2 ♂ and 3 ♀.   Suburbs of Cairo.

2 ♂ and 1 ♀.   Suburbs of Cairo.   Dr. Walter Iunes.

1 ♀.   Mokattam Hills.   Dr. W. Innes.

1 ♂ and 2 ♀.   Gizeh.

2 ♂.   Kafr Amar.

1 ♂.   Fayum.

2 ♂ and 2 ♀.   Tel el Amarna.   Prof. W. M. Flinders Petrie.

2 ♂ and 2 ♀.   Tel el Amarna.

AGAMA MUTABILIS, Merrem.

L'Agame variable ou le Changeant, Is. Geoffr. *op. cit.* pp. 127–129, pl. v. figs. 3 & 4.

Agama mutabilis, Merrem, Tent. Syst. Amph. 1820, p. 50.

Agama inermis, Reuss, Mus. Senckenb. i. 1834, p. 33.

Agama gularis, Reuss, *op. cit.* p. 36.

1 ♂ and 1 ♀.   Desert at Gizeh.   The late V. Ball, Esq., C.B.

4 ♂ and 3 ♀.   Desert at Gizeh.

1 juv.   Desert at Gizeh.   The late Miss R. M. Robertson.

1 ♀.   Abukir.

1 ♂ and ♀ juv.   Mandara.

4 ♂.   Ramleh.

2 ♂ and 1 ♀.   Maryut District.

AGAMA SAVIGNYI, Dum. & Bibr.

13 ♂ and 4 ♀.   From between Suez and Ismailia.   Middle-mass Bey, Inspector-General Coast Guard, Egypt.

1 ♂.   Beltim.   Dr. J. G. Rogers.

1 ♀.   Kafr Amar, below Wasta, on Assiut Railway.

AGAMA SPINOSA, Gray.

Agama spinosa, Gray, Syn. Rept., Griffith's A. K. ix. (1831), p. 57.

Agama colonorum, Rüppell, Neue Wirbelth. 1835, p. 14, pl. iv.

10 ♂, 8 ♀, and 1 juv.   Foot of mountains behind Suakin.

AGAMA STELLIO, Hasselq. & Linn.

Lacerta stellio, Hasselq. & Linn. Iter Palæst. 1757, p. 301.

2 ♂, 2 ♀, and 3 juv. Gabari, Alexandria.

2 juv. Ramleh.

UROMASTIX ÆGYPTIUS, Hasselq. & Linn.

The Dhab or Dab, Shaw, Travels, Barbary & Levant, 1738, p. 250.

The Dab, Bruce's Travels to discover Sources of Nile, v. 1790, p. 198.

Lacerta ægyptia, Hasselq. & Liun. Iter Palæst. 1757, p. 302; Forskål, Descr. An. 1775, p. viii & p. 13; Donndorf, Zool. Beytr. iii. 1798, p. 136.

Stellio spinipes, Daud. Rept. iv. (1803) p. 31; Is. Geoffr. *op. cit.* p. 125, pl. ii. fig. 2.

Uromastix spinipes, Merr. Tent. Syst. Amph. 1820, p. 56.

1 ♀. Between Suez and Ismailia.

1 ♀. Plain of Kafr Gamus.

1 ♂. Beltim. Dr. J. G. Rogers.

1 ♀. Suburbs of Cairo.

UROMASTIX OCELLATUS, Licht.

Uromastix ocellatus, Lichtenstein, Verz. Doubl. Zool. Mus. Berlin, 1823, p. 107; Boulenger, Cat. Liz. B. M. iii. 1887, Corrigenda, p. 499.

Uromastix ornatus, Gray (*not* Heyden), Cat. Liz. B. M. 1845, p. 261.

1 ♀. Neighbourhood of Suakin.

13 ♂, 14 ♀, and 1 juv. Neighbourhood of Suakin.

1 ♀. Wádí Halfa. Major Henry d'Alton Harkness.

VARANUS GRISEUS, Daud.

1 ♀. Suez.

1 ♂. Desert N.E. of Cairo.

1 ♀. Gizeh desert.

1 ♀. Tel el Amarna. Prof. W. M. Flinders Petrie, D.C.L.

1 ♂ and 1 juv. Suakin. Surgeon-Captain R. H. Penton.

2 ♂ and 2 ♀. Suakin.

1 adol. ♀. Tokar, about 50 miles S. of Suakin.

VARANUS NILOTICUS, Hasselq. & Linn.

2 ♂. Luxor.

LATASTIA LONGICAUDATA, Reuss.

Lacerta longicaudata, Reuss, Mus. Senck. i. 1834, p. 29.
Lacerta samharica, Blanf. Zool. Abyss. 1870, p. 449, fig.
Lacerta sturti, Blanf. *op. cit.* p. 452, fig.
Eremias revoili, Vaill. Miss. Révoil aux Pays Çomalis, Rept.
1882, p. 20, pl. iii. fig. 2.
Latastia doriai, Bedriaga, Ann. Mus. Genov. xx. 1884, p. 313.
Latastia samharica, Bedriaga, *l. c.* p. 319.
Latastia longicaudata, Blgr. Cat. Liz. B. M. iii. 1887, p. 55.

4 ♂ and 2 ♀. Suakin. Surgeon-Captain R. H. Penton.
5 ♂ and 2 ♀. Suakin.
1 ♂. Akik, about 80 miles S. of Suakin.
3 ♂, 5 ♀, and 2 juv. Durrur. Colonel A. Hunter, D.S.O.

ACANTHODACTYLUS BOSKIANUS, Daud.

8 ♂ and 5 ♀. Banks of Freshwater Canal, Suez.
4 ♂ and 1 ♀. Abukir.
6 ♂ and 5 ♀. Ramleh.
17 ♂ and 9 ♀. Alexandria (suburbs).
1 ♀. Maryut District.
3 ♂ and 1 ♀. Cairo suburbs. Dr. Walter Innes.
1 ♂. Plain of Kafr Gamus.
13 ♂, 12 ♀, and 2 juv. Margin of desert, Gizeh Pyramids.
9 ♂ and 12 ♀. Plain of Tel el Amarna. Prof. W. Flinders
Petrie, D.C.L.
8 ♂, 4 ♀, and 1 juv. Margin of desert, Luxor.
1 ♀. Oasis of Dâkhel. Major H. S. Lyons, R.E.
6 ♂ and 3 ♀. Assuan.
1 juv. ♀. Suakin. Colonel Sir Charles Holled Smith, C.B.,
K.C.M.G.
3 ♂, 1 ♀, and 1 juv. Suakin. Surgeon-Captain R. H. Penton.
11 ♂ and 11 ♀. Plain of Suakin.
1 juv. Tokar.

ACANTHODACTYLUS PARDALIS, Licht.

16 ♂ and 20 ♀. Maryut District.

ACANTHODACTYLUS SCUTELLATUS, Aud.

1 ♀. Aïn Musa, near Suez.
3. Suez district.
1 ♂. Matariyeh. Dr. Waltèr Innes.
2 ♀ and 1 ♀. Margin of desert, Gizeh. The late V. Ball,
E.-q., C.B.

2 ♂. 4 ♀, and 1 juv. Margin of desert, Gizeh.

1 ♂. Desert north of Birket el Kurun. Major R. H. Brown, R.E.

1 ♂ and 1 ♀. Wádí Halfa. O. Charlton, Esq.

2 ♂ and 3 ♀. Wádí Halfa. Surgeon-Captain R. H. Penton.

EREMIAS MUCRONATA, Blanf.

1 ♀. Plain of Suakin. Colonel Sir Charles Holled Smith, C.B., K.C.M.G.

8 ♂ and 4 ♀. Plain of Suakin. Surgeon-Captain R. H. Penton.

15 ♂ and 16 ♀. Plain of Suakin.

12 ♂, 6 ♀, and 2 juv. Durrur.

EREMIAS GUTTULATA, Licht.

Lacerta guttulata, Licht. Verz. Doubl. Mus. Berl. 1823, p. 101.
Eremias guttulata, Dum. & Bibr. Erpét. Gén. v. 1839, p. 310.

1 ♂. Plain of Suez.

1 ♂. Maryut district.

1 ♂. Plain of Kafr Gamus.

2 ♀. Wádí Hoaf, near Heluan.

1 ♂. Beni Hassan. W. M. Blackden, Esq.

4 ♂ and 4 ♀. Margin of desert, Luxor.

3 ♂ and 3 ♀. Ruins of Karnak.

2 ♂ and 3 ♀. Ruins of Medinet Habu.

2 ♀. Assuan. Major D. F. Lewis.

1 ♂. Philæ.

1 ♂. Suakin. Surgeon-Captain R. H. Penton.

1 ♀. Suakin.

4 ♂ and 4 ♀. Durrur. Colonel A. Hunter, D.S.O.

1 ♂. Erkowit Mountains, west of Suakin.

1 ♂. Akik.

EREMIAS RUBROPUNCTATA, Licht.

Lacerta rubropunctata, Licht. Verz. Doubl. Mus. Berl. 1823, p. 100.
Eremias rubropunctata, Dum. & Bibr. Erpét. Gén. v. 1839, p. 297.

3 ♂ and 3 ♀. Oasis of Khargeh. Professor Sickenberger, Cairo.

1 ♂ and 2 ♀. Plain of Tel el Amarna.

1 ♀. Margin of desert at Gizeh. V. Ball, Esq., C.B., F.R.S.

8 ♂ and 3 ♀. Margin of desert at Gizeh and Abu Roash.

2 ♂ and 1 ♀. Margin of desert at Kafr Gamus.

1 ♂ and 3 ♀. Stony desert plain between Suez Canal and Aïn Musa.

1 ♀. Ras Gharíb. Mr. James Robertson.

MABUIA QUINQUETÆNIATA, Licht.

2 ♂ and 1 ♀. Gardens and roadside, Gabari, Alexandria.

1 ♂ and 1 ♀. Abbasiyeh, near Cairo. Colonel H. .M. L. Rundle, D.S.O.

1 ♂, 6 ♀, and 1 juv. Alluvium, below Gizeh Pyramids.

1 ♀ aud 3. The Fayum.

1 ♂ and 1. The Faỳum. Major R. H. Brown, R.E.

4 ♂ aud 1 ♀. Tel el Amarna.

5 ♂ and 3 ♀. Roadsides, Assiut.

4 ♀. Banks of Nile, Edfu.

2 ♂ and 2 ♀. Banks of Nile, Assuan.

2 ♂ and 1 ♀. Banks of Nile, Philæ.

2 ♂ and 1 ♀. Wádí Halfa. Major Henry d'Alton Harkness.

3 ♂ and 2 ♀. Wádí Halfa. O. Charlton, Esq.

2 ♂, 2 ♀, and 2 juv. Wádí Halfa. Surgeon-Captain R. H. Penton.

1 ♀. Suakin.

MABUIA VITTATA, Olivier.

1 ♂ and 5. Gardens and roadside, Gabari, Alexandria.

1. Fields below Pyramids of Gizeh.

1 ♂, 2 ♀, aud 2 others. Fayum. Major R. H. Brown, R.E.

EUMECES SCHNEIDERI, Daud.

1 ♀. Marsa Matru, about 150 miles to west of Alexandria.

2 ♂ and 5 ♀. Maryut district.

SCINCOPUS FASCIATUS, Peters.

Scincus officinalis, pars, Strauch, Mém. Acad. St. Pétersb. (vii.) iv. no. 7, 1862, p. 41.

Scincus (Scincopus) fasciatus, Peters, Mon. Berl. Ac. 1864, p. 45.

Cyclodus brandtii, Strauch, Bull. Ac. St. Pétersb. 1866, p. 459.

Scincus fasciatus, Boulenger, Cat. Lizards Brit. Mus. iii. 1887, p. 390; Trans. Zool. Soc. xiii. (1891), p. 137.

1 ♀. Suakin. Colonel Sir Charles Holled Smith, C.B., K.C.M.G.

3 ♂, 4 ♀, and 1 juv. Suakin. Surgeon-Captain R. H. Penton.

2 ♂. Suakin.

Known hitherto only from Algeria, Tunisia, and Khartum.

SCINCUS OFFICINALIS, Laur.

1 ♂, 5 ♀, and 3 juv. Desert near Gizch Pyramids.

6. Desert in neighbourhood of Cairo.

CHALCIDES (GONGYLUS) OCELLATUS, Forskål.

2. Marsa Matru.

2. Maryut district.

4. Alexandria.

2. Ramleh.

2. Beltim. Dr. J. G. Rogers.

4. Mahallet el Kebír, Delta. G. H. Keut, Esq.

5. Cairo. Dr. Walter Innes.

7. Gizeh.

4. Fayum.

7. Luxor.

1. Oasis of Khargeh. Professor Ernest Sickenberger.

1. Berys, south of Oasis of Khargeh. Major H. G. Lyons, R.E.

1. Assuan.

3. Philæ.

2. Wádí Halfa.

4. Durrur.

2. Tokar.

CHALCIDES (SPHÆNOPS) SEPOIDES, Aud.

6. Between Suez and Ismailia.

2. Abukir.

36. Pyramids of Gizeh.

1. Kafr Amar.

2. Tel el Amarua.

CHALCIDES (SPHÆNOPS) DELISLII, Lataste.

Allodactylus de l'islei, Lataste, Journ. Zool. v. 1876, p. 238, pl. x.

3. Plain of Suakin. Surgeon-Captain R. H. Pentou.

7. Plain of Suakin.

3. Durrur. Colonel A. Hunter, D.S.O.

20. Durrur.

RHIPTOGLOSSA.

CHAMÆLEON VULGARIS, Daud.

1 ♂ and 2 ♀. Aïn Musa.

2 juv. Egypt.

2 ♂ and 4 ♀. Marsa Matru.

CHAMÆLEON BASILISCUS, Cope.

2 ♂, 1 juv., and 1 juv. ♀. Ramleh.

1 ♀. Tokar. Major H. W. Jackson.

1 ♂, 3 ♀, and 1 juv. Suakin.

1 ♀ and 1 juv. Wádí Halfa. Surgeon-Captain R. H. Penton.

OPHIDIA.

GLAUCONIA CAIRI, Dum. & Bibr.

1. Island of Rhoda, Cairo. Dr. Walter Innes.

1. Luxor, in alluvium.

1. Garden of the Luxor Hotel, among moist grass.

1. Lower floor of a house, Cairo.

1. Durrur, north of Suakin.

ERYX JACULUS, Hasselq. & Linn.

1 ♂. Beltim. Dr. J. G. Rogers.

1 ♂. Mahallet el Kebir. George Kent, Esq.

1 ♀. Abbasiyeh. Colonel H. M. L. Rundle, D.S.O.

2 ♀. Abu Roash.

2 ♀. Gizeh.

1 ♂. Heluan.

1 ♂. Neighbourhood of Cairo.

ERYX THEBAICUS, Reuss.

1 ♂. Fayum. Major R. H. Brown, R.E.

1 ♀. Tel el Amarna. Professor W. M. Fliuders Petrie, D.C.L.

2 ♂ and 2 ♀. Karnak.

3 ♂, 3 ♀, and 1 juv. Suakin.

2 ♀. Tokar.

2 ♀. Durrur.

TROPIDONOTUS TESSELLATUS, Laur.

1 ♀. Beltim. Dr. J. G. Rogers.

ZAMENIS RHODORHACHIS, Jan.

1 ♀. Beni Hassan.

1 juv. Tel el Amarna. Prof. W. M. Flinders Petrie, D.C.L.

ZAMENIS ROGERSII, Anders.
1 ♂. Beltim. Dr. J. G. Rogers.
1 ♀ and 1 juv. Shaluf.
1 ♀. Marsa Matru.
1 ♂. The Desert, Heluan.

ZAMENIS FLORULENTUS, Is. Geoffr.
1 ♀. Beltim. Dr. J. G. Rogers.
1 ♂. Mandara.
2 ♂ and 4 ♀. Abu Roash.
1 ♂ and 1 juv. Gizeh.
1. Fayum.
2 ♂ and 2 juv. Minia.
1 ♀. Tel el Amarna.
1 juv. Assiut.
2 ♂ and 3 juv. Luxor.
1 juv. Karnak.
1 ♀. West bank of Nile, Luxor.
1 ♀ aud 1 juv. Assuan.
1 ♂ and 1 ♀. Wádí Halfa.
1 juv. T'okar.

ZAMENIS NUMMIFER, Reuss.
1 ♂. Beltim.
2 juv. Margin of desert, Heluan. Dr. Adalbert Fényes.

ZAMENIS DIADEMA, Schlegel.
1 juv. East side of Suez Canal, near Suez.
1. Fâyid, west of Bitter Lakes.
1 ♂. Abu Roash.
3 ♂, 1 ♀, and 2 juv. Pyramids of Gizeh.
1 ♂ and 1 ♀. Beni Hassan. M. W. Blackden, Esq.
1 ♂ and 1 ♀. Tel el Amarna. Professor W. M. Flinders
Petrie, D.C.L.
1 ♂. Suakin.
1 ♀. Durrur.

LYTORHYNCHUS DIADEMA, Dum. & Bibr.
1 ♂. West bank of Suez Canal, between Suez and Ismailia.
1 juv. Abu Roash.
1 ♂. Gizeh, margin of desert.
First record of its occurrence in Egypt. Present in the
Sennaar District.

TARBOPHIS DHARA, Forskål.
1 ♀. Beltim. Dr. J. G. Rogers.
1 ♀. Mahallet el Kebir. George Kent, Esq.
4 adol. and juv. Margin of desert, Gizeh.
1 ♀ and 1 juv. Tel el Amarna. Professor Flinders Petrie,
D.C.L.
1 juv. Tel el Amarna.
1 juv. Assuan. Major D. F. Lewis.

CŒLOPELTIS MONSPESSULANA, Hermann.
2 ♂ and 1 ♀. Maryut District.
1 ♂. Alexandria (suburbs).
1 ♂. Mandara. Dr. Walter Innes.

CŒLOPELTIS MOILENSIS, Reuss.
2 ♀. Lower Egypt (? suburbs of Cairo).
1 ♂. Abu Roash.
2 ♀. Suakin.
1 ♀. Durrur.

PSAMMOPHIS SCHOKARI, Forskål.
1 ♀. Aïn Musa.
1 ♂. Between Suez and Ismailia.
1 ♂. Shaluf.
1 ♀. Abbasiyeh. Colonel H. M. L. Rundle, D.S.O
1 ♂. Abu Roash.
3 ♂. Gizeh.
1 ♂. Assuan. Major D. F. Lewis.
1 ♂ and 1 ♀. Suakin Plain. Surgeon-Captain A. H. Penton.
8 ♂ and 4 ♀. Suakin Plain.
1 ♂. Tokar.
1 ♀. Durrur.
1 ♀. Island of Shadwan. Mr. John Strathearn.
1 ♀. Ras Gharíb. Mr. James Robertson.
1 ♂. Berys, S. of oasis of Khargeh. Major H. G. Lyons, R.E.

PSAMMOPHIS SIBILANS, Linn.
1 ♂. Beltim. Dr. J. G. Rogers.
1 ♂. Northern part of Delta. J. R. Gibson, Esq.
1 ♂ and 1 ♀. Mahallet el Kebir. George Kent, Esq.
1 ♂. Abbasiyeh. Colonel H. M. L. Rundle, D.S.O.
3 ♂ and 1 ♀. Abu Roash.
2 ♂ and 2 ♀. Gizeh.

2 ♂ and 1 ♀. Fayum.

2 ♂ and 1 ♀. Minia. Major R. H. Brown, R.E.

1 ♂ and 1 ♀. Tel el Amarna. Professor W. M. Flinders Petrie, D.C.L.

1 ♀. Luxor.

MACROPROTODON CUCULLATUS, Is. Geoffr.

1 ♂. Maryut District.

1 ♀. Ramleh, near Alexandria.

1 ♀. Mandara. Dr. Walter Innes.

1 ♀. Abukir.

NAJA HAJE, Hasselq. & Linn.

1 ♀. Maryut district.

1. Beltim.

1 ♀. Abbasiyeh. Colonel H. M. L. Rundle, D.S.O.

2 ♀. Fields below pyramids of Gizeh, close to water.

1 ♂ and 1 ♀. Fayum. Major R. H. Brown, R.E.

2 ♀. Beni Hassan. M. W. Blackden, Esq.

1 ♂. Tel el Amarna. Professor W. M. Flinders Petrie, D.C.L.

NAJA NIGRICOLLIS, Reichardt.

1 ♀. Assuan.

First record of its occurrence in Egypt.

WALTERINNESIA ÆGYPTIA, Lataste.

1 ♂. Purchased in Cairo from a snake-charmer. Dr. Walter Innes.

CERASTES VIPERA, Hasselq. & Linn.

5 ♂ and 1 ♀. Desert on east side of Suez Canal, between Suez and Ismailia.

1 ♂. Desert on west side of Suez Canal, between Suez and Ismailia.

4 ♂ and 2 ♀. Desert around Cairo.

2 ♂. Desert, Abu Roash.

1 ♂. Desert, Gizeh. R. G. Gallop, Esq.

1 ♂. Desert, Gizeh.

1 ♂. Desert, Beni Hassan. M. W. Blackden, Esq.

1. Ras Gharib. Mr. James Robertson.

CERASTES CORNUTUS, Hasselq. & Linn.

Males with horns.

1 ♂. Assiut.

2 ♂. Suakin. Surgeon-Captain R. H. Penton.

Hornless males.

2 ♂. Luxor.

1 ♂. Wádí Halfa. Major Henry d'Alton Harkness.

1 ♂. Ras Gharíb. Mr. James Robertson.

Females with horns.

1 ♀. Desert east of Suez Canal.

1 ♀. Desert at Gizeh Pyramids.

1 ♀. Desert, Luxor.

3 ♀. Plain of Suakin. Colonel Sir C. Holled Smith, C.B., K.C.M.G.

Hornless females.

5 ♀. Desert, Luxor.

Hornless, sex unknown.

1. Luxor.

1. Wádí Halfa. Major Henry d'Alton Harkness.

1. Ras Gharíb. Mr. James Robertson.

ECHIS CARINATUS, Schneider.

1 ♂ and 1 ♀. Mokattam Hills, Cairo. Dr. Walter Innes.

1 ♀. Assiut.

1 ♂ and 1 ♀. Suakin. Colonel Sir Charles Holled Smith, C.B., K.C.M.G.

1 ♀. Suakin. Surgeon-Captain R. H. Penton.

2 ♂, 2 ♀, and 5 juv. Suakin.

1 ♂ and 1 ♀. Durrur.

BATRACHIA.

ECAUDATA.

RANA MASCARENIENSIS, Dum. & Bibr.

5 ♂ and 5 ♀. Fields below Gizeh Pyramids.

2 ♂ and 1 ♀. Mahallet el Kebír. George Kent, Esq.

4 ♂ and 1 ♀. Freshwater Canal, Suez.

BUFO REGULARIS, Reuss.

2. Ramleh.

1. Beltim. Dr. J. G. Rogers.

3. Mahallet el Kebír. George Kent, Esq.

12. Freshwater Canal, Suez.

4. Canal below Mena.

8. The Fayum. Major R. H. Brown, R.E.

7. Amarna. Professor W. M. Flinders Petrie, D.C.L.
8. Assuan.
2. Wádí Halfa.
1. Wádí Halfa. Surgeon-Captain R. H. Penton.

BUFO VIRIDIS, Laur.

1. In a water conduit, Ramleh.

BUFO PENTONI, Anderson.

2 ♂ and 1 ♀. Shaata Gardens, outside Suakin. Surgeon-Captain R. H. Penton.

17 ♀, 1 ♂, and 1 juv. Wells in Gardens outside Suakin.

SPECIES *said to occur in* EGYPT, *but not observed by me.*

REPTILIA.
CHELONIA.

TESTUDO IBERA, Pallas.

Dr. Keatinge has been so good as to forward to me three photographic views of a land tortoise, three living specimens of which he had purchased for the Museum of the Cairo Medical School.

The species proves to be *Testudo ibera*, Pallas, hitherto known only from North-West Africa, Syria, Asia Minor, Transcaucasia, and Persia. The native from whom they were bought informed Dr. Keatinge that he had got them from the Sudan, and that he had had them alive for more than fourteen years. There is no evidence that this species occurs in Lower Egypt, but, like many others, it may possibly range from Algeria and Tunisia to the Sudan, and, in view of this, I have thought it is as well to record these specimens. If this species is found in the Sudan, it is likely to be distributed in the direction of Wádí Halfa, and even to the plain of Suakin.

SQUAMATA.
LACERTILIA.
GECKONIDÆ.

BUNOPUS BLANFORDII, Strauch.

Mém. Ac. St. Pétersb. (vii. ser.) xxxv. No. 2, 1887, p. 62, figs. 13 & 14.

Egypt: J. Erber, 1870; 2 specimens. Strauch.

GYMNODACTYLUS SCABER, Heyden.

Peters, Mon. Ac. Berl. 1862, p. 271; Gasco, Viaggio in Egitto ',
(pt. ii.) 1876, p. 113.

Egypt: MM. Barnim and Hartmann. Peters.

Near Cairo: Gasco.

Egypt: J. Doubleday. Boulenger.

AGAMIDÆ.

UROMASTIX ACANTHINURUS, Bell.

Nubia: Rüppell, Mus. Senck. iii. 1845, p. 303.

Egypt: A. Duméril, Cat. Méth. Rept. 1851, p. 109.

Egypt: Boettger, Kat. Rept. Mus. Senck. 1893, p. 55.

There is no evidence that this species has ever been found in
Egypt. Rüppell only gives Nubia as the locality whence his
specimens were obtained. One of them went to Paris, where the
locality appears as Egypt; while, on the other hand, the

---

[1] I had long been in search of the late Professor Gasca's work, cited above, as
I was aware that it contained a list of the Reptiles collected, on his journey
through Egypt, in company with the late Prof. P. Panceri. I had, however,
applied in vain to the booksellers for a copy; but on mentioning this to Count
Peracca, he very kindly presented me with one, which enables me to mention
some reptiles which Prof. Gasco collected in Egypt. I think it is evident,
however, that he had no great acquaintance with reptiles, as he refers two
snakes obtained by himself, near Alexandria, to the American genus *Oxyrophus*,
designating them *O. scolopax*, Klein. As some of the species of that genus have
black heads, I am disposed to think that his two specimens were examples
of *Macroprotodon cucullatus*, Is. Geoffr., which occurs in the district of
Alexandria.

It is also stated by Professor Gasco that two examples of *Lacerta ocellata*,
Daud. were obtained in the same locality, and he referred them to a variety which
he called *lepida*. It seems highly improbable, however, that this species should
occur in Egypt, and as Gasco did not distinguish between *Eremias guttulata*,
Licht., and *E. rubropunctata*, Licht., it is just possible that he may have
mistaken an ocellated specimen of the former for *L. ocellata*, Daud. Of course
this is only guess-work, but so unlikely is it that the last-mentioned species
should be found at Alexandria, that I feel compelled to suggest some expla-
nation of how the error may have arisen.

He also records *Psammodromus algirus*, Linn., and says " this species, which
abounds in Algeria and Spain, was collected by us only in the neighbourhood
of Alexandria." My impression is that in this case also we have an error of
identification, and that Gasco had probably before him some species of
Acanthodactylus.

four specimens remaining, iu the Frankfort Museum, have also been referred to Egypt, but why it has been substituted for Nubia is not stated.

SCINCIDÆ.

CHALCIDES (SEPS) TRIDACTYLUS, Laur.

Seps chalcides, Bonap.; Gasco, Viaggio in Egitto, pt. ii. 1876, p. 109.

Neighbourhood of Alexandria: Prof. E. S. Gasco. 2 specimens. This is the only record of the occurrence of this lizard in Egypt. In going through some reptiles in the Cairo Museum I came across one example of this species, but unfortunately there was no information whence it was obtained. As it occurs in Tunisia, it may possibly extend as far east as Alexandria.

RHIPTOGLOSSA.

CHAMÆLEONTIDÆ.

CHAMÆLEON CALYPTRATUS, A. Dum.

Cat. Méthod. Rept. 1851, p. 31; Arch. Mus. vi. 1852, p. 259, pl. xxi. fig. 1.

From the region of the Nile : M. Botta.

OPHIDIA

COLUBRIDÆ.

ZAMENIS DAHLII, Fitz.

Couleuvre, Descr. de l'Égypte, Suppl. Rept. pl. iv. figs. 4. 1 to 4. 3.

Locality unknown. Beyond the fact that the foregoing figure of the species occurs in the ' Description de l'Égypte,' nothing further is known regarding the occurrence of this snake in Egypt. If it is present, it will probably be found in the Delta, possibly in the Maryut district, or between the Suoz Canal and the Nile.

OLIGODON MELANOCEPHALUS, Jan.

F. Müller, Verh. nat. Ges. Basel, vii. 1885, p. 678.

The late F. von Müller has recorded one specimen from Cairo. This species is found in the Sinaitic Peuinsula, so there is nothing remarkable in its presence in Lower Egypt.

i

DASYPELTIS SCABRA, Linn.

Gasco, Viaggio in Egitto, (pt. ii.) 1876, p. 119.

The late Prof. Gasco was the first to record the occurrence of the Egg-eating Snake in Middle Egypt, where he obtained 8 specimens. Count Peracca has been so good as to ascertain from Prof. Costa that two specimens brought back from Egypt by Prof. P. Panceri, the companion of Gasco, one a skeleton and the other in alcohol, are preserved in the Naples Museum.

TARBOPHIS SAVIGNYI, Blgr.

Cat. Snakes B. M. iii. 1896, p. 48.[1]

Couleuvre, Descr. de l'Égypte, Suppl. Rept. pl. iv. figs. 2. 1 to 2. 3.

The remarks I have made regarding *Z. dahlii*, Fitz., apply equally to this species.

VIPERIDÆ.

VIPERA AMMODYTES, Linn.

Linn. Amœn. Acad. i. 1749, p. 506, tab. xvii. fig. 11.
Libya.

This species was recorded by Linnæus from Libya, on the authority of Jonston (Hist. Quadr. et Serp., Lib. ii. 1657, p. 11, tab. i. fig. ammodites), who quoted Solinus as the source of his information.

VIPERA LEBETINA, Linn.

Strauch, Mém. Acad. St. Pétersb. (vii.) xiv. no. 6, 1869, p. 84.
Egypt. Berlin Museum.

BATRACHIA.
ECAUDATA.
HYLIDÆ.

HYLA ARBOREA, Linn.

Hyla savignyi, Audouin, Descr. de l'Égypte, p. 183, Suppl. Rept. pl. ii. figs. 13. 1 & 13. 2.

I have made a most careful search for this species in Lower Egypt, but have never succeded in finding it. It is an analogous case to *Z. dahlii* and *T. savignyi*.

---

[1] I am enabled to make this identification as Mr. Boulenger has favoured me with a sight of the proofs of the third volume of his 'Catalogue of Snakes.'

CAUDATA.

TRITON - — ?
Gervais, Ann. Sc. Nat. (2 sér.) vi. 1836, p. 312.
Oasis of Bahriyeh. A. Lefèvre.
Gervais mentions that M. A. Lefèvre brought a species of
Triton from the oasis of Bahrîyeh.

SALAMANDRA —— ?
Boulenger, Cat. Batr. Grad. 1882, p. 106, footnote.
Near Alexandria. M. Letourneux.
Mr. Boulenger says :—" M. F. Lataste received several larvæ of
a Salamandroid collected near Alexandria by M. Letourneux.
It will probably turn out to be *S. maculosa,* which has recently
been discovered in Syria, and which accordingly will be Circum-
mediterranean."

These are the only two references in zoological literature that
mention the presence in Egypt of this group of Batrachians.

I made a most careful search on two successive years for Salamandroids in the neighbourhood of Ramleh, and on two or three
occasions I employed an intelligent Syrian, who used to collect
for M. Letourneux, to do the same, but neither I nor he ever
succeeded in finding any. I went provided with some British
newts in alcohol to show to the natives, in order to give them
some idea of the kind of animal of which I was in quest, but all
the agricultural labourers to whom I showed them declared that
they had never seen such animals, in the localities I had selected
as appearing to me to be the most likely spots in which to find
Salamandroids. I hope my experience, however, will not deter
others from continuing the search, in view of what has been put
on record by Gervais, and by Boulenger, on the authority of
Lataste.

May 28, 1896.

APPENDIX.

———

I AM indebted to Mr. Boulenger for having directed my attention to an article[1] by Captain P. Parenti and Professor Luigi Picaglia, in which the following species are recorded from Arabia, viz. :— *Hemidactylus coctæi*, D. & B., = *H. flaviviridis*, Rüppell; *Psammosaurus arenarius*, Is. Geoffr., = *Varanus griseus*, Daud.; *Gongylus ocellatus*, Forskål; and *Zamenis florulentus*, Schlegel, = *Zamenis rhodorhachis*, Jan.

The mention of these species necessitates the following additions to the 'List of the Reptiles and Batrachians of Arabia,' in Part V. :—

Page 78. *Hemidactylus flaviviridis*, Rüppell.
Add :—Aden (*Ragazzi*), Parenti and Picaglia, 1886.

Page 79. *Varanus griseus*, Daud.
Add : —Jiddah (*Ragazzi*), Parenti and Picaglia, 1886.

Page 81. *Chalcides* (*Gongylus*) *ocellatus*, Forskål.
Add :—Jiddah (*Ragazzi*), Parenti and Picaglia, 1886.

Page 82. *Zamenis rhodorhachis*, Jan.
Add :—Aden (*Ragazzi*), Parenti and Picaglia, 1886.

The same authors also record that Ragazzi obtained a living Chameleon at Aden, in 1883. They do not give it any specific name, but it was probably *C. calcarifer*, Peters.

———

[1] " Rettili ed Anfibi raccolti da P. Parenti nel viaggio di circumnavigazione della r. corvetta ' Vettor Pisani,' negli anni 1882–85, e da V. Ragazzi sulle coste del mar rosso e dell' America meridionale negli anni 1879–84." Atti Soc. Mod. Mem. (3) v. 1886, pp. 26–96.

INDEX.

k

PRINTED BY TAYLOR AND FRANCIS, RED LION COURT, FLEET STREET, E.C.